Jorge Luis Chinchilla Valverde

Brief historical overview of the concept of number

Jorge Luis Chinchilla Valverde

Brief historical overview of the concept of number

The Importance of the History of Number for Secondary School Teachers

ScienciaScripts

Imprint

Cover image: www.ingimage.com

This book is a translation from the original published under ISBN 978-613-9-43493-0.

Publisher:
Sciencia Scripts
is a trademark of
Dodo Books Indian Ocean Ltd. and OmniScriptum S.R.L publishing group

120 High Road, East Finchley, London, N2 9ED, United Kingdom
Str. Armeneasca 28/1, office 1, Chisinau MD-2012, Republic of Moldova, Europe
Printed at: see last page
ISBN: 978-620-8-33523-6

Contents

Acknowledgements

Alexander, Diana and Claudio, my deepest thanks.
To Edwin Castro, for his contribution, his knowledge. A great friend and a great teacher. Without his support, this work would not have been possible.

Summary

The purpose of this work is to make a brief historical journey through different moments of antiquity, where there was a need to develop strategies and solutions that allowed the people of that time to create or assimilate the notion of number.

This study covers the primitive strategies of man's attempt to count, which could amount to a numerical record from prehistoric times; from the thinking of ancient civilisations such as Mesopotamia and Egypt to the more advanced mathematical developments in Greece and Rome.

We will examine how different cultures have responded to the challenges of quantifying and representing quantities and how their numerical methods and notations have evolved over time. In addition, the impact of these innovations on the development of business, astronomy, architecture and other important areas of daily life and human knowledge will be analysed. It is hoped to offer a comprehensive view of how the digital concept has been essential to the progress of civilisation and how its understanding and use have been crucial to the development of science and technology throughout history. The topic of irrational numbers presents a degree of difficulty in secondary school, where different researchers such as Sirotic & Zazkis (2004) show that this topic is not easily assimilated by students, and even some mathematics teachers lack clarity in the concept of it. For this reason, some very basic notions of some irrational numbers present in the civilisations mentioned above will be briefly presented. Namely:

- Ancient Egypt
- Ancient Babylon
- Ancient Greece.
- Ancient Rome
- Ancient China.

Introduction

In primary and secondary education, the process of teaching numbers is framed within a single numerical knowledge, which seeks a common thread that unifies it and makes it homogeneous. In this respect, in the case of Costa Rica, it is necessary to connect the different sets that are contemplated in the study programmes of Mathematics of the Ministry of Education, in such a way that the students manage to construct, at the end of their process, a knowledge of the set of real numbers.

In this sense, when students enter the initial level, they enter a process that seeks to strengthen the fundamental notions of the set of natural numbers, which is not new to them: children know natural numbers, which they learn in their family or community. In primary school, schoolchildren work with natural numbers in a concrete way in the first years, using objects to count and solve operations; and then they expand their knowledge by generalising and solving written problems about assumed situations.

It is clear that in the natural numbers, given two numbers a and b, their sum: $a+b$, is another natural number and their product: a - b, is also a natural number. However, although in this set two other operations can be carried out, subtraction and division, in some cases the result of subtracting or dividing two natural numbers is not always a natural number.

This leads to the need to study another set of numbers: the integers. This means the need to extend a number system that the students already know. Although the concept of negative numbers is new to them, at their age (11-12 years on average), it is not far from their knowledge either, as they already know situations that allow them to relate to these numbers through contexts that invoke scenarios in their environment, such as the case of specific problems of profit and loss, to cite a classic example. This new set that is formed is called the Set of Integers and is denoted by the symbol Z.

Similarly, from primary school onwards, students intuitively take their first steps in the study of rational numbers when they work with positive fractions and their decimal notation. This allows them to have the previous knowledge for the study of this set, and again through situations associated with their environment, they will be able to establish the elements that make it up, their respective notation and some characteristics that the three sets under study present: N, Z and Q.

For students, assimilating these three sets represents a challenge that they should face in a natural way, associating them with situations common to them. However, this concrete-based tendency presents serious difficulties with the beginning of the study of irrational numbers. The teacher varies the way he teaches this new set of numbers to his students, moving from concrete situations to an abstract concept, in some cases detached from any everyday reality, which generates an epistemological obstacle for the student and therefore a rupture in his perception of the concept of number. Nevertheless, learning this content should enable students to "differentiate" between rational and irrational numbers, which is essential for the construction of the concept of real numbers.

Thus, learning how ancient civilisations such as Mesopotamia and Egypt developed basic number systems provides a solid foundation on the origin and evolution of numbers. This historical perspective helps students to appreciate the importance of the concept of number in their environment and in the development of human knowledge. In addition, learning about mathematical developments in Greece and Rome, where key concepts were formalised, allows students to recognise continuity and progress in the field of mathematics.

By studying the history of numbers, both students and teachers in a classroom can analyse how mathematical ideas have been refined and expanded over time. This historical understanding enriches their learning of integers, rational and even irrational numbers by showing them how these concepts have been used and developed to solve real problems. At the same time, it provides them with a broader perspective and a greater appreciation for the work of earlier mathematicians, fostering a more positive and curious attitude towards mathematics. Knowing the origin and evolution of numbers also helps students to see mathematics not just as a set of abstract rules, but as a living and constantly evolving discipline.

For this reason, the present material focuses on knowing and analysing different historical situations that involve the evolutionary development of the concept of number and in particular, analysing some elements that led to the conformation of the concept and treatment of the irrational number, seen from the perspective of the time and culture shown; all this with the aim of seeking didactic tools that allow the learning of the same.

CHAPTER 1

Why the interest in the history of numbers?

As previously indicated, in secondary school, in general, the teaching of real numbers is carried out around the study of the subsets that make it up: natural numbers, integers, rational and irrational numbers, in order to finally obtain the set of real numbers. This is how, from the classroom, the pupils, through their cognitive construction processes, build up the concept of real numbers in successive moments of their learning.

However, despite the fact that this construction is carried out gradually, from school to the moment of starting with this set of numbers, it is worth asking what degree of understanding the pupils have of the concept of real numbers. As mathematics teachers, personal experience has allowed us to observe that in the best of cases, many of the students manipulate the operations with real numbers in a mechanical way, recite the properties and characteristics of this set, but do not find application and meaning in them. In addition, Zazkis and Sirotic (2010) point out that both students and teachers have problems in recognising rational and irrational numbers. In this sense, both authors indicate that "one of the sources of confusion between rational and irrational numbers is the common use of the rational approximation to the irrational" (p.1).

In addition to the above, Crespo (2008) agrees with this assessment, pointing out that: "many times teachers believe that these numbers have been appropriately constructed, however, indications emerge on occasions that show that irrational numbers are not correctly constructed. The irrationality of some real numbers is a concept that often lacks meaning for students (p. 1)". This leads to the fact that these same students often do not make sense of the work they do in the mathematics classroom. In fact, it is not strange to affirm that the mathematical knowledge presented in schools and educational centres is sometimes not perceived as a social activity, on the contrary, it is presented as complete concepts whose learning path is through the instrumentalisation of memory, the results are taken for granted, leaving aside the needs, the historical construction of them.

For their part, Caicedo & Madrigal (2017) largely attribute the students' reluctance and apathy to traditional teaching methods and the way in which mathematical concepts are taught in the classroom. It can be said that this way of teaching mathematics creates ignorance and confusion among students, so they consider it a "cold" science and only as something indispensable to learn for assessment, because it does not create motivation and meaning in them.

However, this absence of meaning that students have in rational and irrational numbers is likely to increase, for example, many teachers have serious problems in the mastery and management of the concept of irrational numbers. Sirotic and Zazkis (2007) state that a large number of secondary school mathematics teachers lack a clear understanding of what an irrational number actually is, limiting it simply to the particular characteristic of its non-periodic infinite decimal expansion, in contrast to rational numbers. Thus, it is not difficult to think that some students have a weak conceptual basis in the construction of number set structures involving irrational numbers and in the construction of numerical thinking. In addition to meaning, it is also important to look at the structure of numerical thinking, comparisons, estimations, orders of magnitude, etc. The basic aspects are not taken into account in the studies.

It is necessary to emphasise that from an evolutionary, conceptual and historical perspective, the teaching and learning of mathematics is a cultural and anthropological element that includes both the teacher's vision of the human dimension and the interests of people over time; understanding and developing based on what you are interested in learning. Continuing in the case of the set of irrationals, we find in Sirotic and Zazkis (2007) results obtained several decades ago by Arcavi, Bruckheimer and Ben-Zvi (1987), where the fact that the deficiencies of mathematics students and mathematics teachers in the understanding of the irrational number are sometimes surprising is exposed. It is pointed out that there are serious deficiencies in the conceptual part, which generates vague, incoherent and fragmentary ideas. These authors state that the causes that have led to this problem is the presentation of the irrational number in textbooks in connection with very few examples such as the number *n* or the square ratios of the numbers 2 or 3. These weaknesses should make us reflect, analyse and study the teaching and learning of number, in particular the irrational number.

These researchers report several findings on teachers' knowledge, conceptions, and/or misconceptions about irrational numbers. One of the most striking findings of their study is that there is a widespread belief among teachers that irrationality is based on decimals. Sirotic and Zazkis (2007, p.1).

We can therefore emphasise that the above-mentioned drawbacks hinder the understanding, learning and development of numerical thinking, especially thinking related to irrational numbers. All this calls for alternative teaching recommendations, for example those based on historical developments, as well as a critical use of new technological resources for teachers and students. This will allow us to coherently construct the extensions of previously exposed numerical sets, in particular the set of irrational numbers. Learning critically and anaHtically with students can lead to a logical growth and maturity that cannot be achieved by rote learning. Maturity allows to understand the formal concepts of definition of irrationality and at the same time to be aware of the difficulties and problems arising from the historical processes that precisely gave rise to this concept, but above all to adapt some ideas to the school's vision of improving traditional teaching in our classrooms.

It is important to bear in mind that the concept of an irrational number is closely linked to that of a rational number. In fact, it could be said that one way of "approaching" an irrational number is to approximate it by rational numbers. However, the mechanical construction of arithmetical and algebraic contents present in many students limits the possibility of seeing these subjects and mathematics itself as a discipline rich in meaning and sense; on the contrary, they visualise it as a collection of symbols, rules and procedures applied in a mechanical way and lacking in meaning for the learner himself.

This perspective is a product of the type of activities that are developed in the classroom, which do not favour the construction of meanings of the mathematical concepts that are present in the notion of irrational number. Therefore, the absence of meaning mentioned above, sometimes leads to the concept of irrationality being a concept difficult to assimilate and lacking in meaning within a text close to the learner; therefore, careful attention to didactics is essential for an appropriate development of the concept.

This situation makes it essential to create educational contexts that allow us to develop to a greater extent the conceptual meaning that is immersed in the irrational number, which will allow a more solid construction of it. This is what Crespo (2008) states when he indicates:

The concept of real numbers and in particular of irrational numbers cannot be constructed by means of an approach that demands from students only a superficial understanding of some isolated points, such as the assimilation of rules for reading, writing and operations with these numbers.

(p: 28)

However, the lack of educational contexts or teaching strategies that are more meaningful to students is exacerbated when teachers know little about the presence of rational and irrational numbers in different everyday situations, from which the rationale for the subject arises.

Considering the aspects already pointed out, we highlight the use of the history of mathematics as a didactic tool. In this sense, Arcavi, Bruckheimer and Ben-Zvi (1987, cited by Sirotic and Zazkis, 2007) point out that the historical origins of irrationals and their connections with their knowledge will help the teacher, and the student, to understand this set as an activity that is part of the culture of peoples in a changing way according to the beliefs and needs of each moment.

Along the same lines, Miralles & Deulofeu (2005) indicate:

The teacher's own lack of knowledge of the history of mathematics leads to the transmission of mathematics to students as if it were isolated elements, without precedents or subsequent consequences for the progress of scientific knowledge. We understand that it is at this point that knowledge of the history of mathematical concepts can be of great help.

(p: 104)

According to the above, this paper analyses historical aspects of the concept of number, showing elements of the use of rational and irrational numbers in some societies of antiquity, which make evident its necessity and existence. The study of the historical "approach to the evolution" of the concept of rational and irrational numbers will be presented indicating the contributions of mathematicians and ancient civilisations that promoted the study of this concept, with the aim of acquiring a better image of its construction and, from there, to seek strategies that allow us to approach the conceptual structures associated with the historical evolution of the concept of these numbers. The above is a route that can help us as mathematics teachers to rethink those didactic elements offered by the various school textbooks and the associated conceptual schemes of mathematics teachers.

CHAPTER 2

Historical development of the numbering problem

Numbers are so commonplace to us in our daily lives that they cause the deceptive perception that they are inherent, and in adulthood we tend to assume that they have always been present in our minds, like language, just another tool to be learned.

As mathematics teachers, it is necessary to know a little more about the figures that we call numbers and that we use almost innately in our daily context, as indicated above. If we try to go back in time, further back than our Hindu Arabic and Roman numerals existed, it is worth asking: ^Were there other numbers before them? If so^ ^what were they like? We can even speak of Egyptian, Babylonian and even Oriental symbols, but if we strive to go further back in history, we could imagine a beginning where the search for the origin of numbers might reveal some trace of the brilliant invention made by the first human being who conceived the idea of counting.

In this respect, and along the same lines as Ifrah (1987), it is worth asking: (What was the main reason that prompted people to develop the notion of number? Searching for answers to such a question inevitably leads us to a universe of conjectures: was it a necessity to solve an astronomical concern of recording lunar phases, or was it that by observing both hands and both feet they discovered the same number of fingers, or were they even obliged to keep a count of their animals in the form of marks on a bone?

In relation to the above, this same author points out that there are "good reasons" to believe that there was a time when human beings did not know how to count, however this does not imply that they lacked the notion of number, but rather this idea was confined to a kind of numerical sense, that is, to what direct perception allowed them to recognise at a glance. From the above it can be deduced that the concept of number is a concrete reality inseparable from objects and that it manifested itself only in the direct perception of physical plurality.

It follows, therefore, that humans have long been concerned with representing quantities. In this regard, Everret (2021) notes:

Terms for quantities play a pervasive and almost universal role in contemporary human languages, playing a prominent role in the history of the spoken world. Similarly, the numerical focus of human beings is conspicuous in the archaeological record and in the history of writing systems. Numbers are literally etched into our historical record.

(p: 9)

Thus, from the very beginning of human history, mankind felt the need to count his objects, to measure the passage of time, to record the harvest of his crops, to count his animals and to represent real measurements with symbols. Symbols sought to represent numbers and numbers have shaped most cultures. They have transformed human patterns of subsistence over time, made it possible to expand and dominate our environment while allowing the development of other techniques such as agriculture, astronomy and later architecture (Babylonians, Sumerians, Egyptians), essential to human knowledge and inconceivable without numerical speculation.

In this regard, Everett (2021) points to the fact that:

Over the last few decades, archaeologists have uncovered numerous artefacts that show that people in ancient times paid attention to quantities and that they represented them in two dimensions: not in fully developed writing, but in painted marks on cave walls, as well as in engravings on wood and bone. Such counting marks are symbolic in the sense that they refer to something else.

(p: 28)

As a result, different numerical systems were built up. Each culture has developed one or other number systems and symbols to express them, and these have developed throughout history, some of them still exist and others have been lost. Thus, among those we know today, as a product of a long process of study and change, we have complex numbers, imaginary numbers, real numbers, irrational numbers, rational numbers, integers and natural numbers.

In this paper we will comment on the historical evolution of the concept of number. The various archaeological discoveries made to date have revealed various types of manifestations which have been recorded in the form of figurative art, such as burials with mortuary utensils, body adornment, different musical instruments, or marks (linear, punctuated, etc.) intentionally placed on different surfaces, among which we can find the Htica form and, above all, bone marks. It is to this last group of manifestations, often known as "hunting marks", that belong to what are probably the first accounting records in the history of mankind.

It is worth noting that not all the marks left by human beings tens of thousands of years ago are deliberate, nor do all those that are deliberate have any kind of representation of an accounting record. For this reason, we will only focus on showing some archaeological finds that are evidence of deliberate record-keeping.

In addition to the above, Gonzalez et al (2010) point out that it should be taken into consideration that all those palaeohistoric materials that present notches, engravings or drawings do not necessarily have an accounting purpose or use. Thus, only in situations where certain combinations, groupings or specific patterns are observed can we affirm with a certain degree of credibility that we are dealing with an accounting record and, therefore, with a fossil trace of a mathematical thought. Therefore, these authors maintain that [...] "applying this criterion, we will observe that the first relatively evident samples of mathematical thinking belong to rather recent times in the history of mankind, specifically to the Upper Palaeozoic (35,000-10,000 BP) or, at most, to immediately preceding times". Gonzalez et al (2010).

History in mathematics education

If we take into account the history of mathematics in the learning process of this discipline, this will help students to construct mathematical knowledge with more meaning and to get rid of the idea that it is only a didactic content to be fulfilled in a syllabus. In this way, and as Gonzalez (2004) points out, from the point of view of pedagogical effectiveness, not only in the short, but also in the medium and long term, history can contribute to the transmission of knowledge and knowledge to be covered, and to awaken in the student attitudes and methodological habits in accordance with scientific procedures, giving mathematics a practical, pleasant and lively vision, far from what is presented in most of our classrooms, as a dogmatic, closed and finished product.

It is therefore unavoidable to find spaces in our classrooms and learning environments for error, debate of ideas, and understanding of mathematical phenomena. By carrying out these mediations, the teacher is able to mitigate the negative image of what is the Teaching and Learning of Mathematics, considering the integration of the History of Mathematics in its Teaching, seen beyond a simple didactic content, to be of utmost importance.

It is necessary to point out in the first instance that the discussion on the didactic use of the History of Mathematics is not something new. However, Toumasis (1995) cited by Chaves & Salazar (2003) considers that, although much has been said about the need to incorporate the History of Mathematics in Mathematics Education, the material on how to use the History of Mathematics in classroom processes is limited.

Furthermore, Lupiannez (2009) points out that it is necessary to understand that the correct use of history is not an easy task for teachers, as many of them generally lack training in this area, and that it is therefore not easy for students to understand its pedagogical usefulness.

For this reason, the study of the History of Mathematics allows us to know the reason for the origin of different concepts, why they arose and what they were intended to answer. It allows us to know how the different terms and notations that we use today came about. Studying the problems that were posed in another era and how they have evolved, allows the students themselves to make a critical analysis and even deduce just as the ancient mathematicians did.

Considering the aforementioned and what Guzman (2007) states: "knowledge of the history of mathematics should be an indispensable part of the mathematician's knowledge baggage in general and of the teacher at any level of education", this work focuses on analysing historical aspects related to rational and irrational numbers. The aim is for teachers to understand the historical situations that demonstrate the necessity and existence of these numbers, thus facilitating a meaningful construction of these concepts.

Finally, Urbaneja (2004) affirms that there are various motivations for teaching mathematics and, of course, the way of teaching will be positively influenced by this new attitude created by the knowledge of history. In this respect, he gives a very appropriate justification for the objective of this work:

Recreational mathematics is largely based on problems that have been of interest throughout the history of mathematics. It is, therefore, a wellspring of curious problems that can be treated in a playful way as activities outside the classroom and in the framework of complementary cultural activities.

(Urbaneja: 24)

CHAPTER 3

The need to know as mathematics educators about some ancient cultures.

When speaking of the first steps of mathematics, it is necessary to specify that early mathematics required a practical basis for its development, that basis appearing parallel to forms of progress in the society that formulated it. Such a situation appears in cities that were established along great rivers in Africa as well as in Asia, where new forms of collectivities made their appearance: the Nile in Africa, the Tigris and Euphrates in West Asia, the Hindus and then the Ganges in South-Central Asia, and the Hwang Ho and then the Yangtze in East Asia. With swamp drainage, flood control and irrigation, it was possible to turn these lands along the two rivers into rich agricultural regions. Extensive engineering projects were undertaken, which required both funding and administration, necessitating the creation of considerable technical know-how accompanied by mathematics. Thus, it can be said that this early mathematics originated in certain areas of the ancient East primarily as a practical science to aid in agricultural and engineering activities.

These activities required the computation of a useful calendar, the development of systems of weights and measures to serve in the harvesting, storage and distribution of food, the creation of surveying methods for canal and reservoir construction, the partitioning of land, and the evolution of financial and commercial practices to raise and collect taxes and for trade purposes. Thus, the initial emphasis of mathematics was on the practice of arithmetic and measurement. Eves (1964) points out that a special art comes into being by the cultivation, application and instruction of this practical science. We can thus say that the development of human civilisation and the progress of mathematics have gone hand in hand.

If we go back to its beginnings, over thousands of years, mathematicians from many different cultures have created an enormous structure based on numbers. In this sense, the history of mathematics begins with the invention of written symbols to denote these numbers. Without them, civilisation as we know it could not exist. Thus, different number systems were built up. Each culture devised one or another system of numeration and symbols to express them; these developed over the course of history, some have survived and others have been lost. Today we know, as a product of a long process of study and change, complex numbers, imaginary numbers, real numbers, irrational numbers, rational numbers, integers and natural numbers.

An understanding of mathematics in ancient cultures is essential for teachers of mathematics, as it provides a historical context that enriches teaching and highlights the practical evolution of this science. Learning about the development and application of mathematics by civilisations such as the Egyptians, Mesopotamians, Indians and Chinese to solve everyday problems in agriculture, engineering and commerce allows educators to illustrate the relevance and applicability of mathematics in everyday life. This historical knowledge not only facilitates the teaching of fundamental mathematical concepts, but also inspires students by showing them how mathematics has been a crucial tool for human progress over the centuries.

CHAPTER 4

Some Palaeolithic records: the beginning

At this point it is very interesting to note that scholars of this type of Palaeolithic records emphasise a greater interest from a mathematical point of view to those archaeological materials that collect ±30 marks. The original reason seems obvious, since this quantity practically coincides with the number of clias (29.5) of the natural periodical cycle used as a basis for the computation of time by our lunar month predictors. In this respect, the most recent gender studies encourage us to take into account, in addition, that other female menstrual cycle period *of* ±28 days. What has been exposed so far, opens the imagination to think that there may be the possibility of an ancestral evidence of one of the first intellectual activities of man: "the sequential annotation on the basis of a lunar calendar, comprising a period of almost six months". These ideas will be dealt with later.

Following the previous idea, for Gonzalez et al (2010) the first paleotitic piece to which historians of mathematics have made reference is a wolf bone of about 35,000 years ago, found in Vestonice (Moravia, Czech Republic). In this bone, which is about 18 centimetres long, 55 notches can be identified. Its discoverer, Karl Absolom, wrongly interpreted the marks as being grouped five by five and separated by two longer intermediate strokes in two series, one of 30 (= *6x5*) notches, the other of 25 (= *5x5*).

It has subsequently been shown that there is no such five-by-five grouping, yet it seems quite plausible that the carved bone may suggest the accounting record of a series of objects or of some cycle.

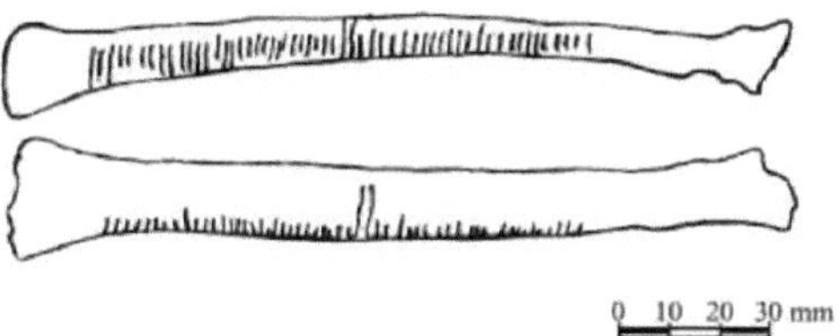

Figure 1: Bone of Dolni Vestonice
Source: Prehistory of Mathematics and the Modern Mind: Mathematical Thinking and Recursion in Franco-Cantabrian PalaeoHtica

For his part, Ifrah (1987) considered that these incisions, in addition to being grouped together, were arranged in two parallel series along two sides of the bone, although he later abandoned this assumption and limited the age of the bone to 20,000 years.

Another archaeological reference is a reindeer antler, found in Brassempouy, France, dated about 15,000 years ago. On this bone, 1, 3, 5, 7 and 9 rectilinear strokes are marked, in a particular arrangement which has given rise to many hypotheses as to the mathematical intentions of its author.

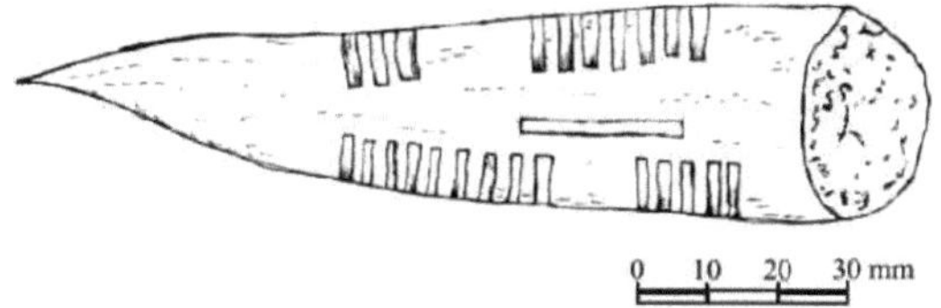

Figure 2: Brassempouy shaft
Source: Prehistory of mathematics and the modern mind: mathematical thinking and recursion in the Franco-Cantabrian PalaeoHithic.

In this respect Ifrah (1987), quoted by Gonzalez et al (2010) believes that this bone is "a kind of arithmetical tool" which presents a graph of the first odd numbers, and even goes beyond that, the marks show a kind of arrangement that allows to quickly find some elementary properties.

These elementary properties, starting from the 2 *x* 2 situation of the marks and taking into account the possible value 1 of the horizontal notch, give us

$$\begin{pmatrix} 3 & 7 \\ 9 & 5 \end{pmatrix}$$

among other possible properties, one has:

$$9 - 7 = 5 - 3 = 2$$

$$3 + 9 = 5 + 7 = 12$$

$$7 - 3 = 9 - 5 = (9 + 5) - (7 + 3) = 4$$

Although this is plausible, Ifrah's (1987) interpretations have not yet been proven.

Along these lines, Steward (2008) points out that early accountants, as he calls them, were already keeping records of "who owned what, and how much", even before writing had been invented, and did not use symbols for numbers. Instead, these accountants made use of clay tokens from about 10,000 years ago in the Near East, which were in the shape of cones, spheres and even eggs.

As time went on, the tokens became more and more elaborate and specialised. Thus some decorated cones represented loaves of bread and diamond-shaped tablets to represent beer. Schmandt-Besserat, quoted by Steward (2008), points out that these tokens represent a first step on the road to numerical symbols, arithmetic and mathematics. These clay marks were simple stripes, "tally marks", which recorded numbers as a series of strokes, such as ||||||||||||| to represent the number 13.

It is therefore necessary to underline the fact that throughout time, the major ideamathematics that has persisted is counting, whether by words or symbols. In addition to this, the international scientific community assumes that counting instruments were first used in Africa 37,000 years ago.

In this regard, Steward (2008) states that the oldest known markings of this type are 29 notches engraved on a baboon leg bone, some 37 000 years old. The bone was found in a cave in the Lebombo Mountains on the border between Swaziland and South Africa.

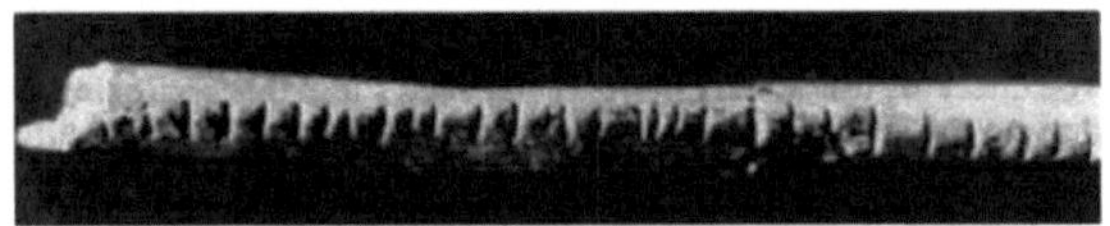

Figure 3: Lebombos bone Source: Taken from the Internet

Another ancient mathematical inscription, the Ishango bone in Zaire, is 25 000 years old. At first glance the markings along the edge of the bone seem almost randomly made, but perhaps there are hidden patterns. One row contains the prime numbers between 10 and 20, namely 11, 13, 7 and 19, the sum of which is 60. Another row contains 9, 11, 19 and 21, which also add up to 60. The third row is reminiscent of a method sometimes used to multiply two numbers by doubling and by repeated division by two. However, the apparent patterns may be mere coincidence, and it has also been suggested that the Ishango bone is a lunar calendar.

It is said that the historical path from the accountants' tokens to modern numerals is long and indirect.

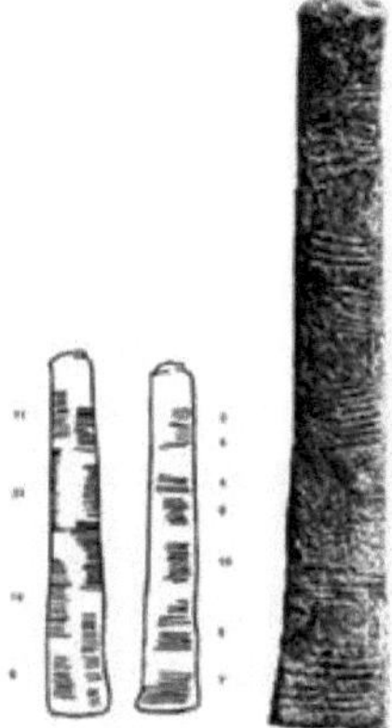

Figure 4: Ishango bone
Source: Taken from the Internet

CHAPTER 5

Mathematics in some ancient peoples

From the shadows of prehistory to the impressive civilisations that emerged some 6000 years ago, the trajectory of mathematics takes us on a fascinating journey through time and the evolution of human thought. As discussed, at the dawn of existence, when primitive communities struggled to understand the world around them, the seeds of incipient mathematics sprouted in an instinctive and elementary way. From the simple need to count to the observation of natural patterns, prehistory saw the first traces of a discipline that would eventually become a universal language for describing reality.

Let us begin a brief overview of ancient mathematics. Pena (2013) states that since approximately 6000 years ago, most civilisations have carried out counting processes as they do today; however, the writing of symbols to represent numbers has been very diverse, and some peoples have even slowed down their scientific and technological advances because they lacked an efficient and agile numerical system.

Thus, it was within ancient civilisations such as the Egyptians and Mesopotamians that mathematics took its first slightly more formal steps. From the banks of the Nile to the fertile lands of Mesopotamia, mankind began to develop more complex numerical systems, to use geometry for urban planning and to apply mathematical concepts to agriculture and astronomy. This period marked a fundamental shift, propelling mathematics forcefully from its practical origins in prehistory into more abstract and systematic terrain, anticipating the astonishing intellectual journey that would continue over the millennia.

Mathematics in ancient Egypt.

The origin of the Egyptian civilisation is not known for sure, but it is certain that it dates back to at least 4000 BC. Unlike the Babylonian civilisation, the Egyptians suffered little outside influence. The Nile River floods their banks annually, but when the waters recede, fertile land remains and was cultivated by the Egyptians. The rest of the country is a desert.

Of the first two thousand years of Egyptian civilisation, between 5000 and 3000 BC, very little is known. Writing had not yet been invented. In the beginning there were two kingdoms: Upper and Lower Egypt. Sometime between 3500 BC and 3000 BC they united. The culmination of Egyptian culture occurs around 2500 BC. At this time the great pyramids were built. In 332 B.C. Alexander the Great conquered Egypt and from then until 600 A.D., Egyptian culture and mathematics belonged to Greek culture.

One of the main aspects that marked the description of mathematics and the complexity of its solution was the type of writing. In fact, throughout Egyptian history, writing evolved little by little, and can thus be divided into three distinct periods:

Hieroglyphic writing: used from 3200 BC to about 2500 BC. In this period the numeration used was similar to that shown in the table below.

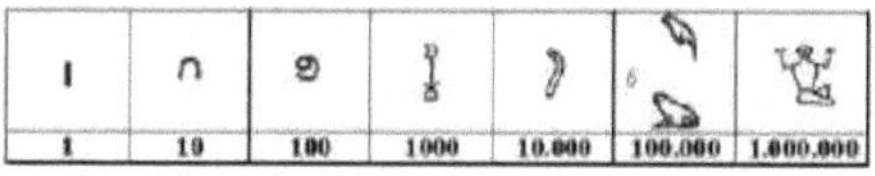

1	10	100	1000	10.000	100.000	1.000.000

Figure 5: Numeration, hieroglyphic writing Source: Taken from the Internet

In particular, the numbers, apart from having to be written with a large number of symbols, represented by objects from the world of rhyme, such as animals, plants, geometric figures, among others, whose line is linear, must be written with a certain aesthetics, grouping the symbols of the same type, if possible in descending order of value. The number of these signs is about 800.

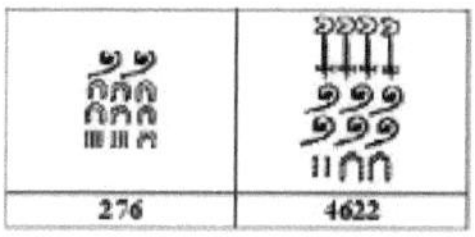

276	4622

Figure 6: Grouping of numbers, hieroglyphic writing
Source: Taken from the Internet

■ **Hieratic script:** used predominantly from 2500 BC to about 600 BC. As before, the numbering used varied with the type of writing, resulting in a greater richness in the number of symbols used to write different digits. It can be said to be an abbreviated system, i.e. a kind of abbreviation of a hieroglyphic sign. For example, instead of using the full figure of a lion, the back of the lion is traced, while retaining the same value or original meaning.

1	2	3	4	5	6	7	8	9

10	20	30	40	50	60	70	80	90

Figure 7: Numeration, hieratic writing Source: Taken from the Internet

■ **Demotic script:** used in the late Late Egyptian period, from 600 BC onwards. This latter variation in writing also produced a slight variation in the symbols used for numbers from the hieratic script.

Today, a very limited part of Egyptian mathematics has remained through inscriptions found in temples and tombs. More information can be found in some of the papyri that are preserved today. Papyrus is a writing medium made from an aquatic plant whose scientific name is *cyperus papyrus* (and which would be the raw material for many other useful things). Papyrus scrolls are

known to have an average length of five metres (the longest papyrus scroll found is over 41 metres long).
The use of papyrus declined at the same time as the old Egyptian culture, being gradually replaced by parchment, of animal origin. It declined throughout the 5th century AD and disappeared completely in the 11th century.

The three papyri

The most important mathematical documents that have survived show exercises posed and solved in three of the most important mathematical papyri so far found: the Rhind papyrus, the Moscow papyrus and the Berlrn papyrus, which date from the hieratic period. There is a widespread misconception that Egyptian science, and in particular mathematics and astronomy, had reached a high level.

The Rhind papyrus, also known as the Ahmes papyrus, owes its name to Henry Rhind, a Scottish Egyptologist who in 1858 acquired a collection of papyri including this one. Dating from 1650 BC, it measures approximately 6 metres long by 33 centimetres wide and its content is purely mathematical with 85 problems posed and solved. It was intended for the training of scribes. Each problem is associated with an aspect of Egyptian daily life, in which operations with fractions are carried out; the areas of the rectangle, the triangle, the trapezium and the circle are calculated (they established that this area is (|*d* j , *d* is the diameter, and that it corresponds to the 9 approximation $n = 3,\ 1605 ...$).

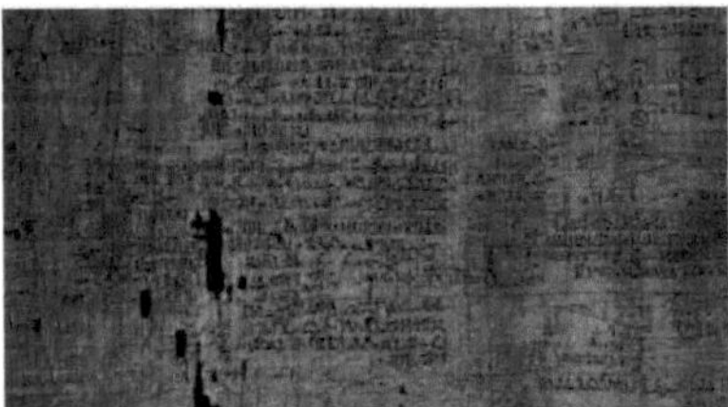

Figure 8: Rhind Papyrus Source: Taken from the Internet

It is considered the oldest Egyptian document dealing with mathematics, and in fact, the oldest mathematical document of any age. Its author is a scribe named Ach-mos'e also known as Ahmes. It begins with the sentence: ^Exact calculation to enter into knowledge of all existing things and of all obscure secrets and mysteries? It is supposed to have the purpose of enlightening the future scribes in the exercise of their activities, (related to collection, distribution, etc.), so that the resolution of the problems is written in a pedagogical way. The five parts of Ahmes' manual deal respectively with arithmetic, stereometry, geometry, the calculation of pyramids and several practical problems.
Furthermore, in the beginning of the papyrus, Ahmes alludes to the fact that the

writing on the papyrus is a compilation of information extini'tii from others 200 years old, which leads to the assumption that the solution of the problems dates back to at least 1900 BC.

Beckmann (2006) quotes the following paragraph (translation) with which he begins the ancient text in question:

Careful calculation. The entrance to the knowledge of all things that exist and all dark secrets. This book was faithfully copied in the year 33, fourth month of the season of the flood under the King of Upper and Lower Egypt, A-user-Re, in the enjoyment of life, from an ancient writing made in the time of the king of Upper and Lower Egypt, Ne-mat'et-Re....

(Beckmann: 30)

Among the problems posed are those related to multiplication and division, unit fractions, areas of rectangles, triangles and circles (approximation of *n*), resolution of equations with 1 unknown and calculation of volumes and calculations on pyramids.

The Moscow Papyrus: originally known as the Golenishchev Papyrus. Author unknown. Orthographic and palaeographic studies have made it possible to date it to the 13th dynasty (ca. 1759-1630 BC). The Moscow Papyrus, so called because it is deposited in the Moscow Museum of Fine Arts, is 5 m long and 8 cm wide. It consists of 25 problems, some of which are too damaged to be read. It is written in hieratic (a simplification of hieroglyphic) and dates from around 1890 BC, from the time of the 12th dynasty. The scribe, whose name we do not even know, was not as good at his craft as Ahmes, whose writing and way of explaining things is much clearer... The image below shows the original papyrus in hieratic script at the top and its translation into hieroglyphic script at the bottom.

Figure 9: Moscow Papyrus Source: Taken from the Internet

From this papyrus we can highlight the problems related to areas of rectangles and triangles, volumes of truncated pyramids, calculation of the surface area of a "basket", linear equations and unit fractions.

The Berlm Papyrus: is a collection of mathematical and medical papyri dated around 1300 BC. Like the Moscow papyrus, the author of the itself is unknown.

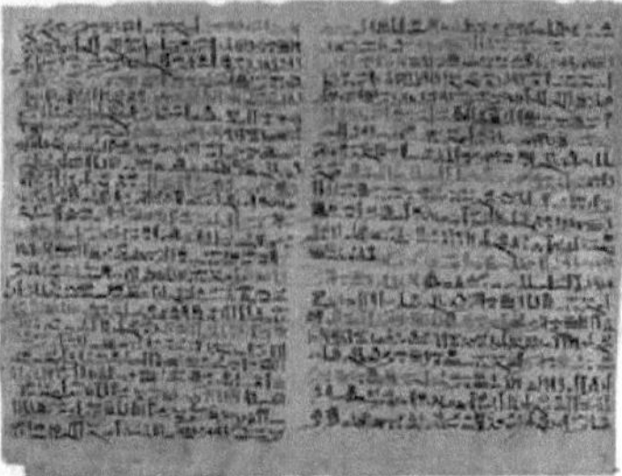

Figure 10: Berlm Papyrus Source: Taken from the Internet

It contains problems related to unit fractions, linear equations and a system of two equations with two unknowns (one of which is also of second degree).

Egyptian Arithmetic

For Morales (2002) and as mentioned above, we cannot really speak of a single system of numeration, because they used two: the hieroglyphic system using hieroglyphs and the sacral (sacred) system used by the priests using cursive symbols, which in the 8th century BC led to the demotic system or system of the people, also cursive and in an abbreviated form. According to Berciano (2007), the first characteristic to highlight is that thanks to the complete knowledge of the tables of duplication and the calculation of the thirds of a number, the scribes handled with complete ease the four elementary operations: addition, subtraction, multiplication and division.

His hieroglyphic system was decimal but not positional, in which the additive principle establishes the arrangement of the symbols. The use of this principle allows any number to be expressed, each symbol being repeated as many times as necessary. They used different symbols for the units, tens, hundreds, etc. Thus, the Egyptians had pictograms to represent 1, 10, 100, 1,000, 10,000, 100,*000*, 100,000 and 1,000,000. The 1 was represented by a vertical stroke , the 10 by a horseshoe, the 100 by a coiled rope or a kind of spiral, the 1,000 by a lotus flower (including its stem), the 10,000 by a raised and twisted finger, the 100,000 by a tadpole, and the 1,000,000 by a kneeling man with raised arms, apparently representing the god Heh, god of infinity and eternity, holding up the sky.

	LECTURA DE DERECHA A IZQUIERDA	LECTURA DE IZQUIERDA A DERECHA
1		
10		
100		
1.000		
10.000		
100.000		
1.000.000		

Figure 11: Fundamental ciphers of the Egyptian hieroglyphic numeration
Source: Universal History of Figures, Ifrah. G

Thus, a number like 34 was written in *П П П П* |||| . This system is not suitable for the large calculations demanded by astronomy, although as mentioned above, this system of numeration was additive, so that performing a sum was relatively easy due to the simple accumulation of equal numbers or pictograms and regrouping them. For example, if there were ten vertical lines (1), they were replaced by a horseshoe (10), a spiral (100) and so on.
For example, the sum of 2,816 and 348 would be:

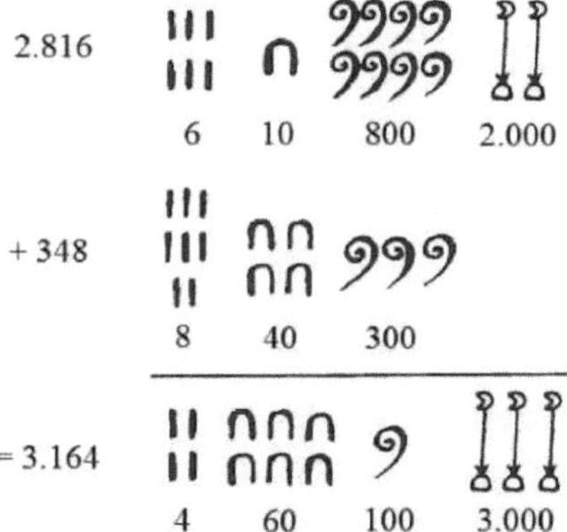

Figure 12: Sum of figures Source: Taken from the Internet

Moreover, the use of unit fractions, i.e. fractions whose numerator equals the unit, except for 2/3 and 3/4, is a feature that almost immediately distinguishes the body of mathematics developed in Ancient Egypt. Their hieroglyphic writing represented schematically a mouth, followed by the numerical value of the denominator:
Thus^ we have the following example:
It is curious that the sacred parts of the Eye of Horus, where the denominators

are powers of two, were also used to refer to the fractions of hekat (measure of grain).

Figure 13: Sum of figures Source: Taken from the Internet

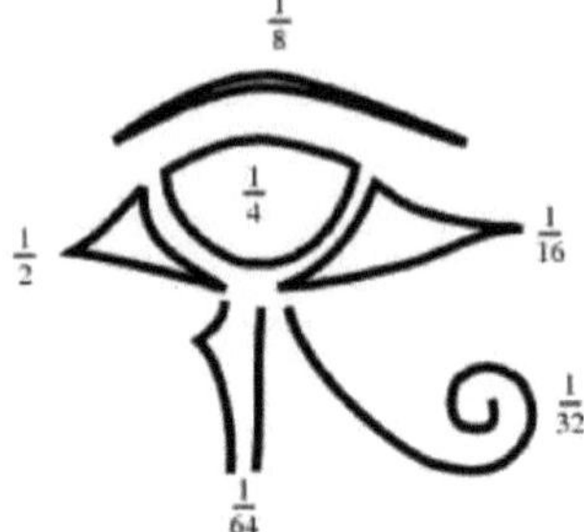

Figure 14: Eye of Horus Source: Taken from the Internet

However, Corrales (2022) states that there is only evidence that the Egyptians used fractions of denominator 1, with the exception of the fractions $\frac{2}{3}$ y $\frac{3}{4}$, , which had their own symbolism, which was a little different from the other fractions. When they wanted to express fractions with different numerators, they had to express them as sums of fractions with numerator 1:

$$\frac{2}{5} = \frac{1}{2} + \frac{1}{15}$$

For his part, Moreno (2012) indicates that to facilitate calculations, the ancient Egyptians prepared tables in which the most common fractions were divided into sums of unit parts. The Ahmes Papyrus begins with a table of factorisation of all irreducible fractions with a numerator of 2 and a denominator of 5 to 101.

2/5	1/3+1/15	2/53	1/30+1/318+1/795
2/7	1/4+1/28	2/55	1/30+1/330
2/9	1/6+1/18	2/57	1/38+1/114
2/11	1/6+1/66	2/59	1/36+1/236+1/531
2/13	1/8+1/52+1/104	2/61	1/40+1/244+1/488+1/610
2/15	1/10-1/30	2/63	1/42+1/126
2/17	1/12+1/51+1/68	2/65	1/39+1/195
2/19	1/12+1/76+1/114	2/67	1/401/335+1/536
2/21	1/14+1/42	2/69	1/46+1/138
2/23	1/12+1/276	2/71	1/40+1/568+1/710
2/25	1/15+1/75	2/73	1/60+1/219+1/292+1/365
2/27	1/18+1/54	2/75	1/50+1/150
2/29	1/24+1/58+1/174+1/232	2/77	1/44+1/308
2/31	1/20+1/124+1/155	2/79	1/60+1/237+1/316+1/790
2/33	1/22+1/66	2/81	1/54+1/162
2/35	1/30-1/42	2/83	1/60+1/332+1/415+1/498
2/37	1/24+1/111+1/296	2/85	1/51+1/255
2/39	1/26+1/78	2/87	1/58+1/174

2/41	1/24+1/246+1/328	2/89	1/60+1/356+1/534+1/890
2/43	1/42+1/86+1/129+1/301	2/91	1/70+1/130
2/45	1/30-1/90	2/93	1/62+1/186
2/47	1/30+1/141+1/470	2/95	1/60+1/380+1/570
2/49	1/28+1/196	2/97	1/56+1/679+1/776
2/51	1/34+1/102	2/99	1/66+1/198
		2/101	1/101+1/202+1/303+1/606

Figure 15: Table of fractions Ahmes papyrus
Source: Taken from the Internet

Surfaces, areas and the number *n*

The Egyptians devoted a great deal of effort to the calculation of areas as they were primarily an agricultural society. Because after the annual rise of the Nile, each person had to be re-assigned the same area of land as before the flood. This fact gave rise to the need to know how to calculate the area of different surfaces and, depending on the type, we find different exercises posed and solved. In fact, the various problems dealt with in the aforementioned papyri, based on the volumes and areas of the most familiar plane and solid figures, were for the most part correctly solved.

Romero et al (2021) point to the fact that when asked about the three most important achievements of Egyptian geometry, there is a general consensus about two: "the approximations of the area of the circle" and "the deduction of the rule to calculate the volume of a pyramid trunk", but there is some disagreement about the third: "did they indeed try to find the correct formula for the area of the surface of a hemisphere?

In this context, Gervan (2015) presents an analysis of problem 10 of the Moscow papyrus, where it is unfortunately in a fragmentary state, even to the point of losing certain written parts that are crucial to understanding its nature. It is divided into three columns with a total of 14 lines, in the sixth of which the damage is unavoidable. The author therefore proposes a transcription and transliteration that leaves aside any attempt to reconstruct this small lost fragment.

In this respect, the same author quotes Struve[I] , who makes a very interesting analysis based on a re-reading of problem 10 of the Moscow Mathematical Papyrus.

[I]STRUVE, V., Mathematischer Papyrus der..., op. cit., pl. 10

Figure 16: Problem 10 Moscow Papyrus

Source: Mathematical Practice in Ancient Egypt. A rereading of Problem 10 of the Moscow Mathematical Papyrus.

Here is the transcription with some changes taken from the material of Romero (2021) in order to better understand the problem:

1. Example of the calculation [i.e. the calculation of the area] of a basket.
2. They give you a basket with a mouth [i.e. an opening, i.e. diameter].
3. of 4+1/2 at the edge ([presumably, in diameter]), joh!
4. Let me know [the value of] your area. Haras

Suggested solution

5. Take 1/9 of 9, because basket
6. is half of an egg (i.e. a hemisphere):, which becomes [i.e. results] in 1
7. You will make a difference, which is 8.
8. Haras 1/9de8
9. which becomes 2/3 + 1/6 + 1/18.
10. You will make up the difference of this with 8 after
11. (subtract) 2/3 + 1/6 + 1/18, which becomes 7 + 1/9.
12. Haras [the multiplication] 7 + 1/9 by 4 + 1/2
13. Look, [this] is its area [lit. of it].
14. jYou found it well!

Romero et al (2021) propose the following resolution reformulating the problem as follows:

Find the surface area of a hemisphere of diameter 4 1/2.

The solution that this author suggests can be expressed symbolically as follows:

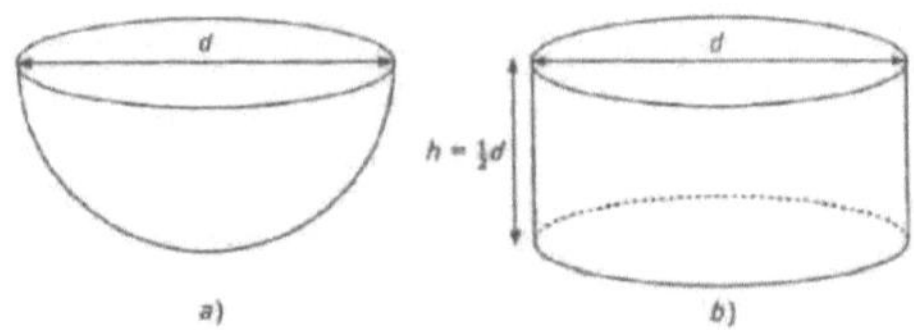

Figure 17: Problem 10 Moscow Papyrus

Source: Analysis of algebraic equations from Egyptian culture to the present day.

$$A = 2d(8/9)(8/9)d = 2d^2(8/9)^2 = 2\pi r^2$$

where the Egyptian value of n is 256/81 (**value to be discussed later**). It is clear that such an expression is identical to the present formula for the curved surface of a hemisphere $(A = 1/2\pi d^2)$ with a different value for n. This should draw our attention to the ingenuity and ability of the Egyptian mathematicians of the time. In fact, these authors indicate that

If this depiction of the Egyptian method is accurate, then we are looking at an even more remarkable achievement than the application of the correct formula for the volume of a pyramid stem, for the very idea of a curved surface (not simply a surface that can be obtained by rolling a flat surface) is a very advanced mathematical concept. This would anticipate the innovative work of Archimedes (ca. 250 BC) by some fifteen hundred years.

(Romero et al: 13)

However, it is important to clarify that doubts have been raised about the interpretation of the term "egg". Peet (1931) quoted by Romero et al (2021) has argued that this term should be interpreted as a semi-cylinder, in which case the above suggested solution, expressed symbolically, (Figure 17, b) would become $A = 2\pi h,$, where $\pi = 256/81$ y $h = 1/2d$ is the height. This would be the Egyptian counterpart of the modern formula for the area of the curved surface of a semi-cylinder. In problems 42, 43 and 44 of the Rhind papyrus the explicit calculation of the area of a circle is posed, and in problem 50 it is solved for a circle of diameter 9 units. For this, given the arguments that appear in the original papyrus, the logical reasoning followed by the scribe would be the following:

As a first step we consider the square circumscribed to the circle. Its side measures 9 units, whose area is equal to 92 = 81 units2.	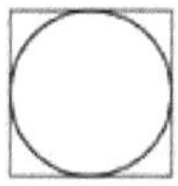
This square is subdivided into squares of unit side, giving a grid of 9 × 9 unit squares.	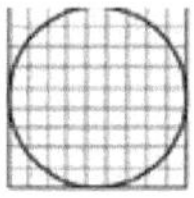
The diagonal of the squares of side 3 units of the corners is traced, obtaining an irregular octagon, which will be taken as an approximation of the circle.	

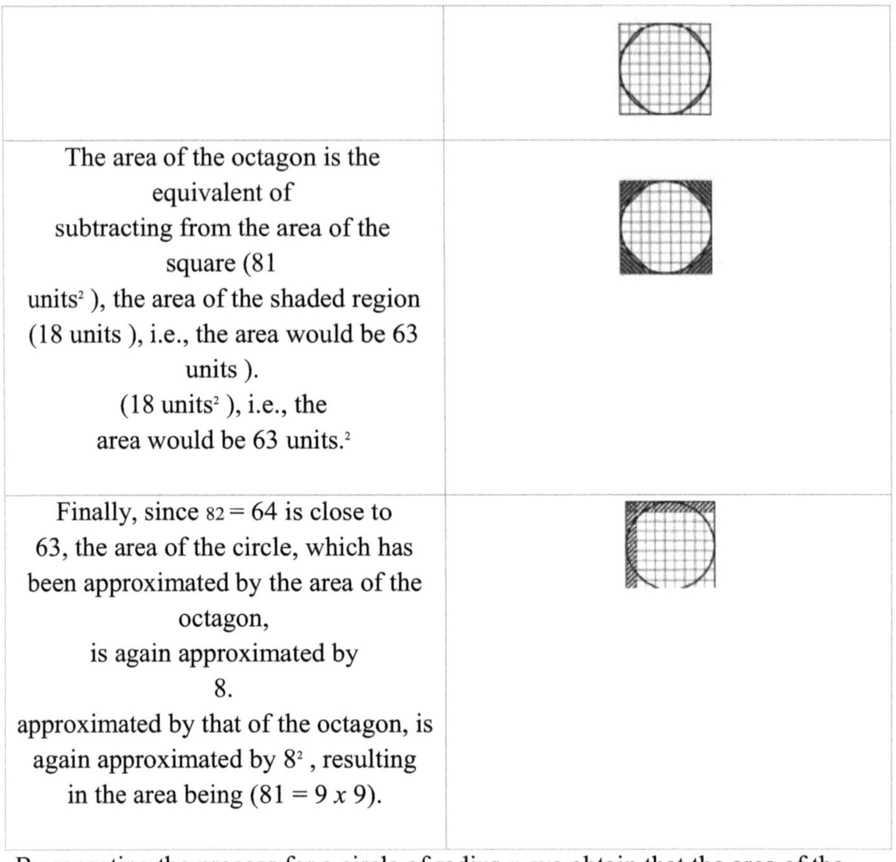

The area of the octagon is the equivalent of subtracting from the area of the square (81 units^2), the area of the shaded region (18 units), i.e., the area would be 63 units). (18 units^2), i.e., the area would be 63 units.^{2}	
Finally, since 82 = 64 is close to 63, the area of the circle, which has been approximated by the area of the octagon, is again approximated by 8. approximated by that of the octagon, is again approximated by 8^2 , resulting in the area being $(81 = 9 x 9)$.	

By repeating the process for a circle of radius *r*, we obtain that the area of the circle

$$\text{Area} = \left(\frac{8}{9}d\right)^2 = \frac{64}{81}d^2 = \frac{256}{81}r^2$$

Therefore, the number *n*, although not mentioned in any papyrus expHcitly, is approximated by the fraction

$$\pi = \frac{256}{81} \approx 3,16049$$

In this respect, Beckmann (2006) states that "It is clear that Ahmes cheats twice: first by stating that the area of the octagon is equal to that of the circle, and then by taking 63" 64. However, it is worth noting that both approaches are similar, although not entirely so. (p. 31).

Pythagoras Triangle in Ancient Egypt

There is evidence that the ancient Egyptians used a special type of triangle in many of their constructions, drawings and paintings, the so-called "sacred triangle", a rectangular triangle with sides in a 3 - 4 - 5 ratio, to which they attributed magical or aesthetic qualities.

However, one of the doubts that persists over the centuries is whether in ancient Egypt the "Pythagoras theorem" was known for any right-angled triangle of any sides a and b. The ratio 3 - 4 - 5 allowed them to be able to delimit surfaces with right angles, used for example in the construction of the pyramids. The relation 3 - 4 - 5 allowed them to be able to delimit surfaces with right angles, used for example in the construction of the pyramids.

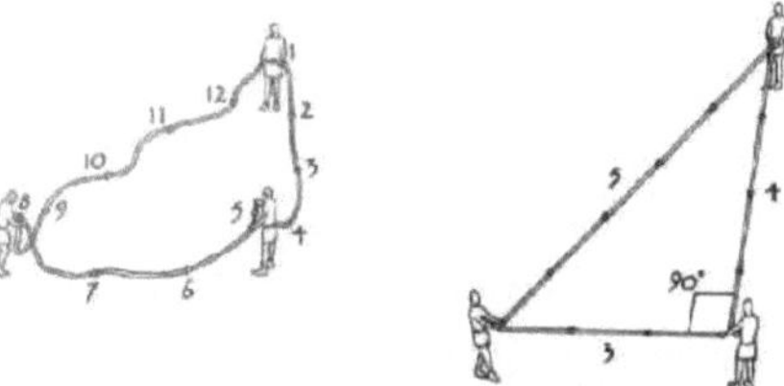

Figure 18: Rope with 13 knots.
Source: Workshop: Ancient Egyptian Mathematics May, 2019

Let's look at the method in question. First of all, they took a rope and tied 13 knots, equidistant two by two. By joining the two ends they obtained a right triangle with sides 3, 4 and 5, and therefore the right angle they were looking for.

Figure 19: Rope with 13 knots.
Source: Taken from the Internet

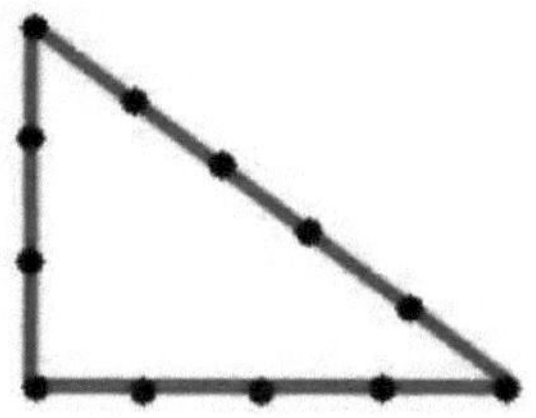

Figure 20: Triangle 3,4, 5; Rope with 13 knots.
Source: Taken from the Internet

In fact, the interior angles of such triangles are approximately $36^\circ\ 52'\ 11''$ and $53^\circ\ 07'\ 48''$. Therefore, any triangle possessing these angles will have sides in the above proportion and will be a "Sacred Triangle".

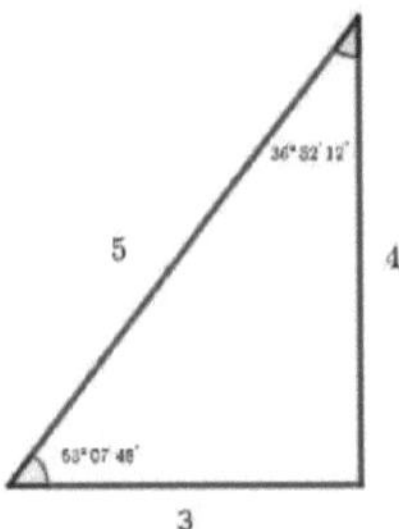

Figure 21: Sacred Triangle Source: Own elaboration

In this regard, several Egyptologists agree that one of the best examples of the use of the sacred triangle can be found in the royal pyramids built during the 6th Dynasty. Martinez (2001) indicates that pyramids designed with sacred triangles contain four such triangles in their structure and are formed with each apothem of the pyramid, its base and its height. At present, no documents have been found to prove that the ancient Egyptians knew the Pi- tagoras theorem, however it should be noted that in order to solve the sacred triangles, it is not necessary to use this theorem, since they can solve with basic additions or additions, without using squared numbers or solving complicated square raffles.

In relation to the above, Sanchez (2012) indicates that Herodotus, among other chroniclers, records its use when referring to the work of surveyors tracking earth movements caused by the flooding of the Nile. In terms of architecture, the use of the sacred triangle is well documented, for example in the construction of the Great Pyramid of Kefren in the 26th century BC. Exposed references to the Pythagorean relationship appear in a few specific numerical examples in Egypt, although no document has survived that sets it out in a general way. For example, a document from the Twelfth Dynasty (ca. 2000 BC) found at Kahun uses the following terminology:

$$1^2 + \left(\frac{3}{4}\right)^2 = \left(1 + \frac{1}{4}\right)^2$$

which is proportional to that of the Egyptian triangle.

Square Rafces in Ancient Egypt

For Gillings (1982), squaring rational numbers was something that can be found frequently in various ancient papyri, however, when talking about square roots, it is less common, although they were expressed, not calculated. In fact, the ancient Egyptians were not required to formulate a method for finding the square roots of perfect squares. It is likely that they used a table that was drawn up by the scribes and composed of whole numbers of perfect squares. These would have been complete enough for all their usual needs. There is also evidence that this culture worked with a system of two equations and two

unknowns. Thus, the Berlm Papyrus contains two problems where one involves a quadratic equation in two variables. It is known that the Egyptians did not indicate the resolution of these quadratic equations, but there is evidence showing their use, such as the problem of the squares. We find the following problem: *You are told that the area of a square of 100 square cubits is equal to the sum of the area of 2 more squares. The side of one of them is* 1/2+1/4 *of the other. Find the sides of the squares.*
In our present context, this problem is transformed into solving the system of equations:

$$x^2 + y^2 = 100$$

$$y = (1/2 + 1/4)x$$

with x, y the sides of the squares sought.
Romero et al (2021) summarise the solution by means of Egyptian rhetorical algebra of this problem given by the scribe:
Take a square of side 1 cubit (that is, a false value of y equal to 1 cubit). Then, the other square will have a side of 1/2 + 1/4 cubits (that is, $x = 1/2 + 1/4$). The areas of the squares are 1 and 1/2 + 1/16 square cubits, respectively. Adding the areas of the two squares gives 1 + 1/2 + 1/6 square cubits. Extract the square nn'z from this sum: 1+1/4. Divide this 10 by 1+ 1/4, which gives 8 cubits, the side of a square. (Thus^ from the false assumption of $y = 1$, we have deduced that $y = 8$.) At this point the papyrus is so badly damaged that the rest of the solution has to be reconstructed. One can only assume that the side of the smaller square was calculated as 1/2 + 1/4 of the side of the larger square, which was 8 cubits. Therefore, the side of the smaller square is 6 cubits.

(Romero et al: 5)

The Babylonians

I personally believe that every mathematics teacher should know a little about Babylonian culture. Thus, in understanding the historical trajectory from countable engravings to modern numerals, we see that their path was long and broken. Over thousands of years, the peoples of Mesopotamia developed agriculture, moving from a nomadic way of life to various settlements in a kind of city-states: Babylon, Eridus, Lagash, Sumer, Ur.
The Babylonian civilisation encompasses a group of peoples (Sumerians, Akkadians, Chaldeans, Assyrians, Babylonians and others) who lived in Mesopotamia, the region between the Tigris and Euphrates rivers, which today is the Republic of Iraq. Around 1900 BC, the Babylonian civilisation appeared.
Babylon was the cultural centre of the so-called Fertile Crescent between about 2000 and 500 BC, which led to the evolution of mathematics; it was not affected

by the invasion made by the Persians commanded by Cyrus in 538 BC.

However, the Babylonians were not the first inhabitants of this region, for they were preceded by the Sumerians, around 3500 BC, and later by the Akkadians, around 2300 BC. In this cultural exchange, the Babylonian people managed to assimilate most of the cultural legacy of both civilisations, at the same time increasing and enriching them.

It is now known that their primitive symbols were inscribed on clay tablets and then transformed into pictograms (symbols representing words by means of simplified images of what they wish to express), and were later reduced to pictograms, being reduced to a small number of cradle-shaped marks, impressed into the clay by the use of a dry stylus with a flat, sharp end. Thus, by 3,000 BC the Sumerians had developed an elaborate form of writing, now called cuneiform: "cradle-shaped".

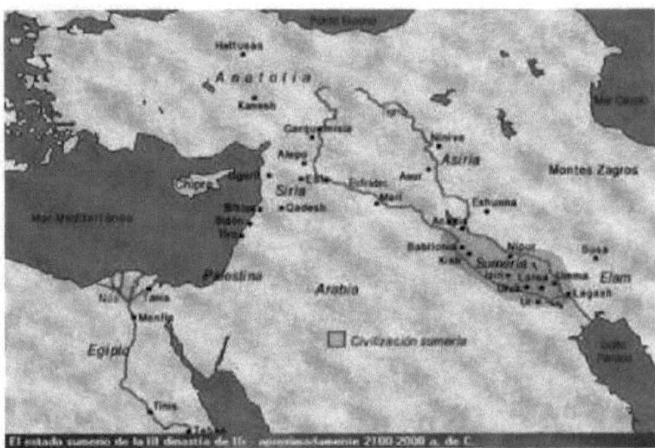

Figure 22: Map of ancient Mesopotamia Source: Taken from the Internet

It was not until the end of the 19th century that archaeologists began to excavate in ancient Mesopotamia, where numerous finds were discovered over many years. Up to the present day, approximately half a million clay tablets have been collected, 300 are related to mathematics and two hundred have engraved tables of multiplication, division, squares, cubes, roundels and compound interest.

Some of the clay tablets found that contain mathematical texts come from the late Sumerian period (2100 BC), others belong to the early Babylonian dynasty (King Hammurabi, 1700 BC) and others date from the period of Nebuchadnezzar to the Seleucid period (600 BC - 300 AD). Although the first tablets with mathematical texts were found, and the presence of numerical texts was detected, they could not be deciphered until the 1930s, thanks to the work of Grotefend and Rawlinson.

Returning to the cuneiform script, Steward (2008) reports the existence of two different types of cradle: a thin, vertical cradle to represent the number 1: Γ and a thick horizontal cradle for the number 10:^{4} . These cradles are divided into groups to indicate the numbers 2-9 and 20-50. However, this pattern stops at 59, and the thin cradle then takes on a second meaning, the number 60.

This is believed to be the reason for the Babylonian number system being base 60, or sexagesimal. That is, the value of a symbol can be a number, or 60 times that number, or 60 times 60 times that number, depending on the position of the symbol. This situation is very similar to our decimal system, in which the value of a symbol is multiplied by 10, or by 100, or by 1,000, depending on its position.

In this respect, if we compare this system with ours, we will notice that we not only use ten symbols to represent arbitrarily large numbers: we also use the same symbols to represent arbitrarily small numbers. We use the sign called "decimal point". Thus, when we place the digits on the left side of the decimal point, they represent whole numbers, and if we place them on the right side of the decimal point, they represent fractions.

In fact, these special fractions are the multiplications of a tenth, a hundredth and so on. Thus, the figure 36.47 indicates 3 tens + 6 units + 4 tenths + 7 hundredths. This same situation was already known to the Babylonians and they used it with an extraordinary result in their astronomical observations. Different studies point to the Babylonian equivalent of the decimal point by a semicolon (;), but this is a "sexagesimal point" and the multiplies to its right are multiplies of 1/60, (1/60 x 1/60) = 1/3600 and so on. As an example, the list of numbers 12, 59; 57, 17 means 12x60+59+ 57/60 + 17/3600.

As mentioned above, the Babylonian representation of numbers was by means of wedges (cuneiform); thus the number 1 was represented by a single wedge, and the number 10 by a kind of arrow pointing to the left. The Babylonian representation of the numbers from 1 to 59 is obtained by adding up to 5 arrows and 9 wedges.

Figure 23: Babylonian Representation of Numbers Source: Taken from the Internet

Fractions were dealt with very briefly earlier when some results from the ancient

Egyptian period were presented. In the case of the Babylonians, they first used them about 4000 years ago. Their numbering system, as indicated, was sexagesimal, and all their fractions were used with that starting point. Moreover, this fractional system allowed them to make very accurate approximations of rafces. Corrales (2022) states that the fractional notation system used by the Babylonians was the best that had been developed until the Renaissance.

Note that the number sixty has many divisors, so the use of the sexagesimal base greatly facilitated operations with fractions.

In this respect, Ortiz (2005) comments that the so-called **Akkadian** system, whose numerical symbols are very similar to those of the previous table, is the highest arithmetical development of the Babylonians and used positional notation. In arithmetic, some fractions were given unique symbols. For example,

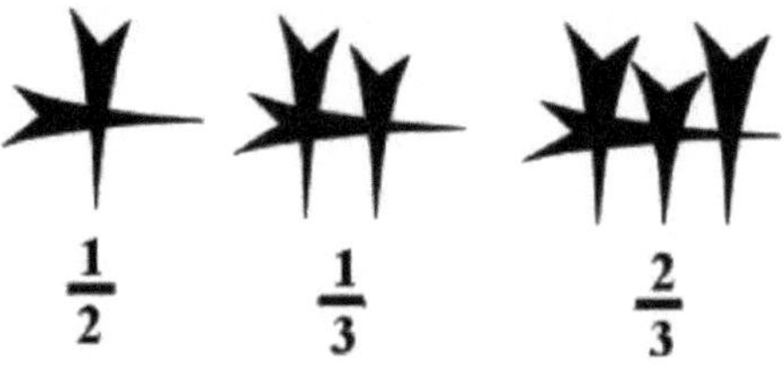

Figure 24: Babylonian representation of fractions Source: History of Mathematics

It can be seen that the notation was based simply on the inclined arrangement of one of the symbols. An example of his way of working his numbering system in base 60 is given by Fernandez (2010) who makes a comparison with our base 10, for a better understanding. So^ for example we see that in base ten, each number takes a different meaning depending on their positions:

$$7235,25 = 7 \cdot 10^3 + 2 \cdot 10^2 + 3 \cdot 19 + 5 \cdot 10^0 + 2 \cdot 10^{-1} + 5 \cdot 10^{-2}$$

Now, to write numbers in sexagesimal notation, follow certain criteria, however each criterion is not standardised for representing numbers in sexagesimal notation and varies according to the author or working group that elaborates it:

- Separate each of the positions by a low dot.
- Use the comma to indicate that the position of the number is less than a whole number.

Then, following a scheme similar to the number in decimal base, we can place the digits of a number in base sixty in boxes:

$$1,2,31,24,6,36 = 1 \cdot 60^3 + 2 \cdot 60^2 + 31 \cdot 60^1 + 24 \cdot 60^0 + 6 \cdot 60^{-1} + 36 \cdot 60^{-2}$$

Thus we have:

603	602	601	600	60^{-1}	60^{-2}
1	2	31	24	6	36

Now, you will remember how to convert the number 7235.25 from base 10 to base 60.

Whole part

1. Integer division: Divide 7235 by 60 and write down the quotient and the remainder.

$7235 \div 60 = 120$ (quotient) with a residual of 35

2. Repeat the process with the quotient obtained until the quotient is 0.

$120 \div 60 = 2$ (quotient) with a residual of 0

$2 \div 60 = 0$ (quotient) with a residual of 2

Therefore, the remainders of the divisions in reverse order give us the integer part in base 60: 2,0,35.

Fractional Part

For the fractional part, 0.25:

1. Multiply by 60:

$$0{,}25 \times 60 = 15$$

As we have reached a whole number, we end aqrn.

Therefore, the fractional part in base 60 is 15.

Final Result

We put the whole part and the fractional part together:

7235.25 in base 10 is 2.0.35, 15 in base 60.

This can be read as 2 units of 60^2 , 0 units of 60^1 , 35 units of 60^0 , and 15 units of 60 .$^{-1}$

It should be noted that the Babylonians did not use any symbol to represent the number zero, that is, they did not have a symbol to express the *vatphic space*, which led to certain ambiguities in the representation of certain numbers. According to Ortiz (2005), instead of zero, the Babylonians used a white space (of a certain amplitude). For example, the symbols < *p* can signify the number 11, the number 11-60, the number < p, the number < p, the number < p, the number < p, the number < p, the number < p, the number < p, the number < p.

11 - 602,.... Later, for the place of zero they used the symbol .

For example, , $7424_{(60)} = 2 \cdot 60^2 + 3 \cdot 60^1 + 33$

In addition to the above, it should be noted that the Babylonian culture was characterised by being the cradle of great mathematicians, who were able to perform arithmetical operations with ease, despite not having an algorithm for division; they also solved equations of second, third and fourth degree, as well

as systems of equations. They calculated sums of arithmetical progressions, some geometrical ones and even worked with sequences of squares.
It is said that the Babylonian scribes were experts in the matter of calculation, the key to their success lay in the tablets, for they were able to learn by heart large lists of calculations so that they could apply them to problems of all kinds.
The Babylonian mathematical texts mostly specialised in the handling of many problems for the appropriation of concepts, at no time was the general method or any demonstration of a theorem enunciated. It was enough to present hundreds of examples in which only the coefficients of the problem were varied, which were usually about the handling of series of numbers and geometric relations.
In geometry, the Babylonians knew the properties of similar triangles, moreover, they could calculate the length of the diagonal of a rectangle, which indicates that they knew the Pythagorean triads and the square rafces.

Algebra in ancient Babylonia

In the tablets found with mathematical content related to algebra, there are a large number of problems that show the knowledge obtained by the Babylonians regarding the solution of first and second degree equations.
In this sense, according to Mayoral (2009) the equation of second degree was often posed as the solution of two equations with two unknowns, of the form:

$$x + y = a$$

$$x \cdot y = b$$

where x, y, are obviously the solutions of the quadratic equation: $z^2 - za + b = 0$
Note that if $y = a - x$ and substituting in $x - y = b$ we have:

$$x(a - x) = b \Leftrightarrow x^2 - ax + b = 0$$

The square root in ancient Babylonia

The Babylonian knowledge of geometry is highlighted:
1. The use of the Pythagorean theorem
2. Calculating the area of simple figures.

The Pythagorean Theorem used by the Babylonians..: The use of the Pythagorean Theorem dates back to 1700 B.C., i.e. some 1200 years before the birth of Pythagoras. A translation of a Babylonian clay tablet preserved in the British Museum reads as follows:

- 4 is the length and 5 is the diagonal. What is the width? Its size is unknown. 4 times 4 is 16.
- 5 times 5 is 25.
- Subtract 16 from 25 and 9 remain.
- How many times how much should I take to get 9?

- 3 times 3 is 9.
- 3 is the width.

In addition, on Yale University's YBC Tablet 7289, dating from about 1600 BC, there is a square drawn and the resulting rectangular triangles drawn on the diagonals.

Figure 25: Babylonian tablet representative of the square root of 2
Source: Taken from the Internet

As can be seen, it is interesting to note that the Babylonians studied the number $\sqrt{2}$. This number was represented in the form . The Babylonian tables (YBC 7289) (c. 2000-1650 BC) give an approximation of the square root of two in four sexagesimal digits (The sexagesimal system is a positional number system using the digit sixty as a base) which is analogous to six decimal places:

$$1+\frac{24}{60}+\frac{51}{60^2}+\frac{10}{60^3}=1,41421296$$

It is noteworthy that this approximation of $\sqrt{2}$ is far superior to that obtained by the Greeks much later.

On the other hand, the Plimptom tablet is the most important mathematical document of ancient Babylonia. It is dated between 1900 and 1600 BC and has been described by several historians, the interpretation given in 1945 by Neugebauer and Sachs in their book Mathematical Cunei- form Texts being very significant. The Plimpton tablet appears to be a simple record of accounts of commercial transactions, but interpreters have claimed to see a description of Pythagorean numerals and even primitive trigonometric tables.

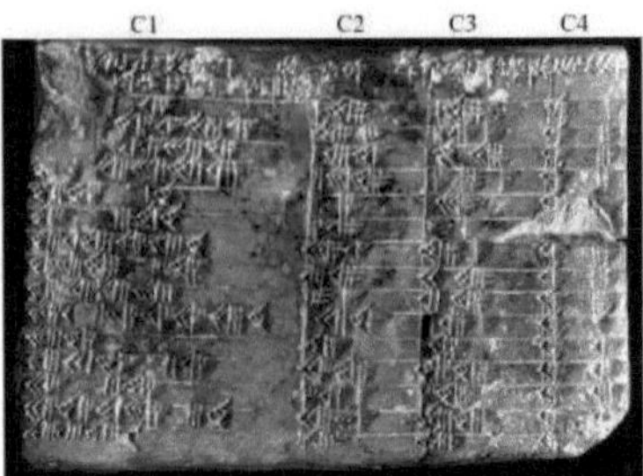

Figure 26: Plimptom tablet Source: Taken from the Internet

Fernandez (2010) states that the initial decipherment of the tablet corresponds to Neugebauer and Sach, who published it in Mat-hematical Cuneiform Text, in 1945. It is a tablet with 4 columns (C 1 - C 4) by 15 rows (plus a header). To proceed to the interpretation consider the following right triangle, which represents geometrically the tern pitag(a, b, c):

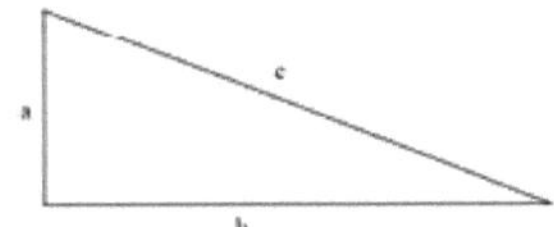

Figure 27: Interpretation Plimptom tablet 1
Source: Own
Elaboration

- C 4: The last column is labelled with an order number (from to 15) for each row.
- C 2: One of the legs appears (e.g. "a ").
- C 3: Refers to the hypotenuse "c".
- Cl: Refers to the expression (c/b)2.

$(c/b)^2$	a	c	C4
1,59.0.15	1.59	2.49	1
1,56.56.58.14.50.6.15	56.7	1.20.25	2
1,55.7.41.15.33.45	1.16.41	1.50.49	3
1,53.10.29.32.52.16	3.31.49	5.9.1	4
1,48.54.1.40	1.5	1.37	5
1,47.6.41.40	5.19	8.1	6
1,43.11.56.28.26.40	38.11	59.1	7
1,41.33.59.3.45	13.19	20.49	8
1,38.33.36.36	8.1	12.49	9
1,35.10.2.28.27.24.26	1.22.41	2.16.1	10
1,33.45	45	1.15	11
1,29.21.54.2.15	27.59	48.49	12
1,27.0.3.45	2.41	4.49	13
1,25.48.51.35.6.40	29.31	53.49	14
1,23.13.46.40	56	1.46	15

Figure 28: Plimptom 2 tablet interpretation Source: Own elaboration

The value of each leg "b" is determined by the relevant operation between the first and third columns. It can be seen that the ratio c/b corresponds to the cosecant, so this tablet is considered to be an early and incipient form of trigonometry, without unnecessarily exaggerating this circumstance as the birth of this mathematical discipline.

Making the change to decimal notation (see table below) and calculating the angles, we see that they correspond to angles ranging from about *45°* to about *58°*.

a	b	c	α
119	120	169	45°11'10,704"
3367	3456	4825	45°58'47,186"
4601	4800	6649	46°17'0,738"
12709	13500	18541	46°58'17,707"
65	72	97	47°73'59,804"
319	360	481	48°36'26,355"
2291	2700	3541	49°54'46,935"
799	960	1249	50°18'16,167"
481	600	769	51°22'33,656"
4961	6480	8161	52°45'1,546"
45	60	75	53°10'24,491"
1679	2400	2929	55°1'55,225"
161	240	289	56°11'35,875"
1771	2700	3229	56°50'2,844"
56	90	106	58°8'44,199"

Figure 29: Plimptom 3 tablet interpretation Source: Own elaboration

Mathematics in Ancient Greece: Today's history assigns to mathematics the character of a deductive science, the undisputed legacy of the Greek civilisation of the Classical period, from about 600 BC to 300 BC. The best example of this mathematics is found in Euclid's Elements (approximately 325-265 BC), a work that reveals part of the progress achieved by this civilisation during the aforementioned period, and which constituted for two millennia the model of mathematical work, with its distinctive features of axiomatisation and the use of demonstration as a means of validation of established mathematical statements (or theorems). Moreover, for the so-called classical Greeks the study of mathematics was indispensable for the development of the intellectual capacities of man, in particular, of the statesman.

This fact is important to rescue, because for other ancient cultures such as the Chinese or Hindu culture, to name a few, these did not consider mathematics as a fundamental knowledge, bringing as a consequence a stagnation in all the sciences and techniques of these cultures.

For the classical Greek culture, due to the founding influence of Pythagoras of Samos

(569-475 BC, approx.) and his school, mathematics was regarded as the science suitable for the intellectual interpretation of the world as a rational, orderly and harmonious system, not chaotic, capricious and arbitrary.

Therefore, it can be said that this mathematical development was driven by the Pythagorean philosophical point of view, in which everything in nature was expressible in terms of positive integers and relations between them, i.e. everything is summed up in numbers.

However, this particularity suffers a strong setback with the discovery of

irrational numbers, which provokes the first great crisis in mathematics. This situation unleashed in the Greeks a quest to absorb this blow to their Pythagorean philosophy, which led to the development of a greater mathematical knowledge on the part of several thinkers who were concerned with adapting the idea of irrationality to the mathematical theory of the time.
This is stated by Gartia (2009) with regard to incommensurability:

The question of incommensurability was taken out of the domain of Greek number theory or arithmetic and into the realm of geometry. In the latter branch they tackled the question by elaborating a theory of proportions, including incommensurable magnitudes, which gave a full mathematical account of the subject; a work attributed by historians to Eudoxus of Cnidos (ca. 390 BC - ca. 337 BC) and embodied by Euclid in Book X of his famous work.

(Gartia: 62)

It can be said that the relevance of the development of such a theory arose from the discovery that A/2 was not the only number of its kind, in addition, A/3, A/5, A/6, A/7, A/8, A/10, A/11, A/12, A/13, A/14, A/15 and A/17 were irrational as they appear in some Greek texts.
Plato (428 BC-347 BC) is known to have put in the words of Theaetetus:

Theodore taught us some calculation of the rafces of numbers, showing us that those of three and five are not commensurable in length with that of one, and then continued in this way up to that of seventeen and seven, on which he stopped. Judging, then, that the raffles were infinite in number, it came into our minds to try to understand them under one name, which suits them all.

(Platon: 9)

Along the same lines, at the time Euclid published his Elements, at the beginning of the Alexandrian or Hellenistic period (around 300 BC-600 AD), the theory of incommensurable magnitudes was already part of the mathematical knowledge of that time and was accepted by the Greek mathematical school, which was reflected in his best known works: Elements and Data, where he develops the theory of the irrationals. Thus, in the Elements he makes a study of most of the results concerning irrationals that were already known at that time.
This historical epoch begins with the conquests of Alexander the Great, who came to form a great empire with a newly founded capital in Egypt: Alexandria, and who intentionally unified Greek culture with that of the Near East.
As a result, the theoretical conception of the classical Greeks blended with the practical vision of the Babylonians and Egyptians to produce a new mathematical thought, which, despite having a more geometrical (theoretical) pragmatism, did not neglect the practical applications of its theoretical findings.
For his part, Garda (2009) points to examples of this new attitude in

Alexandrian mathematics in the works of the Greek mathematicians Archimedes of Syracuse (ca. 287-212 BC) and Apollonius of Perga (ca. 262- ca.200 BC). Same attitude found in those who sought, in the classical style, to study nature mathematically and who in doing so faced the resolution of problems that required the operative handling of quantitative data, as was the case with the astronomers Hipparchus of Nicaea (ca. 180- ca. 125 BC) and Claudius Ptolemy (ca. 85- ca. 165 AD). This situation leads to the fact that, when applying mathematics, the problem arose of calculating quantitative data that did not correspond to rational numbers, as was the case for *n*, so that in practice one had to be satisfied with approximations calculated with techniques derived from theoretical knowledge and not from empirical knowledge. Thus it is that among these scenarios there were approximations of square ratios of numbers that were not perfect squares, that is, irrational, and the example par excellence is the method devised by Heron of Alexandria (ca. 10 - ca. 75 A.D.), which will be discussed in this paper.

In this work it was preferred to leave aside the study of Euclid's Elements in detail although in many parts Euclid's results and his working tools are mentioned and used.

Heron of Alexandria

Figure 30: Heron of Alexandria
Source: Taken from the Internet

Heron of Alexandria was probably born around the year 10 in Alexandria (Egypt), where he carried out most of his work, and died around the year 75 also in Egypt. He wrote at least 13 works (some of them are clearly textbooks) on mechanics, mathematics and physics in general. He invented several mechanical instruments, most of them for practical use, designing more than 100 machines with different utilities.

However, he is best known as a mathematician, both in the field of geometry (calculation of areas of triangles, quadrilaterals and regular poHgons and areas and volumes of prisms, pyramids, cylinders, cones and spheres) and in geodesy (the branch of mathematics that deals with the determination of the size and configuration of the Earth).

His main treatise on measurement, Metrica, consists of three books, and in all of

them the theoretical support for the formulas used is given, so that they are not mere enumerations of rules to be used; moreover, in his examples he does not use measurements but numbers without units of measurement. In Book I Heron presents a formula, mathematically demonstrated, for calculating the area of a triangle from its three sides, which today bears his name although it is known to be due to Archimedes. The so-called Heron formula is $A^2 = s(s-a)(s-b)(s-c)$ where s is the semi-perimeter, i.e. $s = (a+b+c)/2$.
A It is important to note that this formula prevails today and can be found in all geometry textbooks used in secondary school.
Moreover, being a formula applicable to any triangle, the result of the multiplication did not always have to be a perfect square, but could not be. Therefore, to obtain the area of the triangle A it is necessary to calculate the square root of the result of the said multiplication, and in the case that this was not a perfect square, Heron in his book proposes a technique to approximate the square nn'z. This technique will be studied later, basically because it is an accessible method for secondary school students, its geometric interpretation and its high degree of approximation.
Mathematics in China: Throughout history, Chinese mathematics has evolved autonomously, without direct influence from other civilisations. This is largely due to China's geographical location, the communication barriers of the time, and the particular way in which the Chinese integrated aspects of foreign cultures during invasions. In contrast to Greek mathematics, there is no axiomatic development. The concept of mathematical proof in China is very different from that of the Greeks, but the approach and the results obtained are nevertheless surprising. In fact, it is not easy to establish the beginnings of its origin.
According to the Chinese historian Ling Wang, the beginnings of Chinese mathematics date back to the 16th century BC. In this development, the decimal system, hieroglyphic numbers and special calculating devices, knots, boards, squares, squares, compasses, compasses were always used.
In ancient China, as in Egypt and Babylon, mathematical books consisted of compilations of practical problems, in which the problem, the answer and sometimes the method of arriving at the solution were presented. Despite the discovery in 1984 of the Suan shu shu (A Book of Arithmetic), a text dated around 180 BC, our knowledge of Chinese mathematics before 100 BC is very limited. It was found near Jiangling, in Hubei province, and is written on bamboo strips. Today hardly any traces remain of texts written during the Han dynasty, but Emperor Shih Huang-ti of the Chin dynasty ordered them to be

burned in 213 B.C. Later, during the following Han dynasty, mathematicians were forced to rewrite the ancient books. Among the most prominent works are the Zhou bi suan jing (The Classic of Gnomon Arithmetic, cited above, and the Circular Paths of Heaven), sometimes known as Chou Pei Suan Ching, and the Chiu chang suan shi (The Nine Chapters on the Mathematical Art), also called Jiu zhang suan shu.

The exposition in them is dogmatic: the conditions of the problem are formulated and the answers are given. After exposing a series of problems of the same type, a solution algorithm consisting of a general rule is given, however, the deduction of this rule does not appear.

In this respect, "Mathematics in Nine Chapters" was reworked throughout the Middle Ages, this book became a mathematical encyclopaedia not only in China but also in neighbouring countries such as Vietnam and parts of India.

According to Rftnikov (1987) the value of $n = 3$ is used in the first chapter and seems to come from very ancient times. However, Chinese mathematicians of that time were able to calculate the value of n more accurately. Thus, for example, for the 1st century of our time, in the works of Lin Sing one can find the value of $n = 3{,}1547$, in the 2nd century, in the manuscripts of ChisanHen, we have that n = д/Ю- For the 3rd century, calculating the value of the sides of inscribed poHgons, Liu Hui found that $n = 3{,}14$, for this, he started from the idea that the area of the circle is approximated by default by the areas of the inscribed poHgons. For the case of the approximation by excess to the area of these poHgons, the areas of the rectangles circumscribed around the remaining ones are added.

segments. In the 5th century Tsu Chung-Chih found for n two values of convenient fractions $: \frac{22}{7}$ y $\frac{385}{113}$, giving an approximation for the value of n to a seventh 7113 digit: $3{,}1415926 < \pi < 3{,}1415927$.

The following problems, taken from the work mentioned above, are given below: Problem 36 (Chapter I). Given a field in the form of a circular segment of base $78\frac{1}{2}$ and height $13\frac{7}{9}$ find the area.

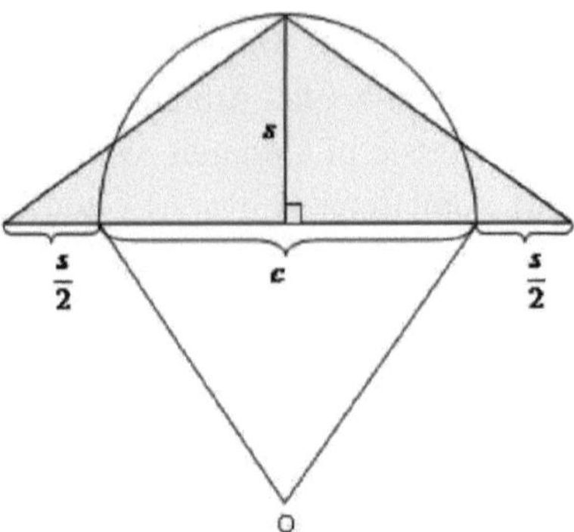

Figure 31: Sagita Source: Own elaboration

The height was formerly called Sagittarius (Sagittarius) and the base is called the chord. For the area of a circular segment as a function of y the Chinese used the formula:

$$A = s\frac{c+s}{2}$$

The exact formula for calculating the area of the circular segment with modern resources can be expressed as follows

$$A = \left[\arccos\left(\frac{c^2-4s^2}{4s^2+c^2}\right)\right]\left(\frac{4s^2+c^2}{8s}\right)^2 - \frac{c\left(\frac{4s^2+c^2}{8s}-s\right)}{2} =$$
$$\left(\arcsin\frac{4cs}{4s^2+c^2}\right)\left(\frac{4s^2+c^2}{8s}\right)^2 - \frac{c\left(\frac{4s^2+c^2}{8s}-s\right)}{2}$$

Problem $N\circ$ 11 (Chapter IV). Determine the dimensions of a door knowing the diagonal and the difference between the length and the width. With the current notation the problem is reduced to two equations:

$$x^2+y^2=c^2 \qquad x-y=k$$

Which is written as:

$$2x^2+2kx+k^2-c^2=0$$

In the text it is written:

$$X_{1,2} = \sqrt{\frac{c^2-2(\frac{k}{2})^2}{2}} \pm \frac{k}{2}$$

In the treatise there are no deductions or demonstrations of the assumption that the above values were obtained in the following way:

$$X_{1,2} = z \pm \frac{k}{2}$$
$$X_1^2 + X_2^2 = 2z^2 + 2\frac{k}{2}^2 = c^2$$

From where:

$$Z = \sqrt{\frac{c^2 - 2(\frac{k}{2})^2}{2}}$$

Problem $N°$ 15 (Chapter IV). Given a square field of area 5677522 (What is the length of the field: In the fourth book called Shao-Huan or for other authors: Distribution by progressions, problems involving irrational numbers are proposed and solved, for example:

1. Extraction of square and cubic roots.
2. Given a square of given area calculate the side.
3. Calculation of the radius of a circle given its area.

The methods used by the Chinese to extract roots are similar to those learned many years ago by students in the first years of secondary school.

It is known that the ancient Chinese were well acquainted with the Pythagorean theorem. In the last chapter (Gou Gu) of the Jiuzhang suanshu, as previously indicated, twenty-four problems related to the properties of right triangles are shown. In ancient China, the base of a right triangle was known as *kou or gou*, the height *ku or gu*, and the hypotenuse *hsian oxian*.

Liu Hui (ca. 3rd century AD) was a Chinese mathematician who made an important mathematical breakthrough in a commentary on the Jiuzhang suanshu or Nine Chapters of the Mathematical Art around 263. In particular his commentary on the problem presented in Gou Gu, describes the solution to the right triangle problem by means of a diagram and becomes evidence that ancient China recognised the Pythagorean theorem. In it, Liu uses the Pythagorean theorem to calculate the height of objects and the distance to those objects that cannot be measured directly. It is important to mention that the diagram includes a central yellow square, which is not mentioned in Liu Hui's commentary on the demonstration of *the gou-gu* relationship. Despite this, the diagram is still relevant, as we shall see below. First, look at one of the triangles formed by drawing the diagonal of a rectangle; the shorter side of the triangle is the gou side, and the square on that side is coloured in red. Similarly, the longest side of the rectangle corresponds to the *gu* side of the triangle, and the square on that side is coloured in blue-green.

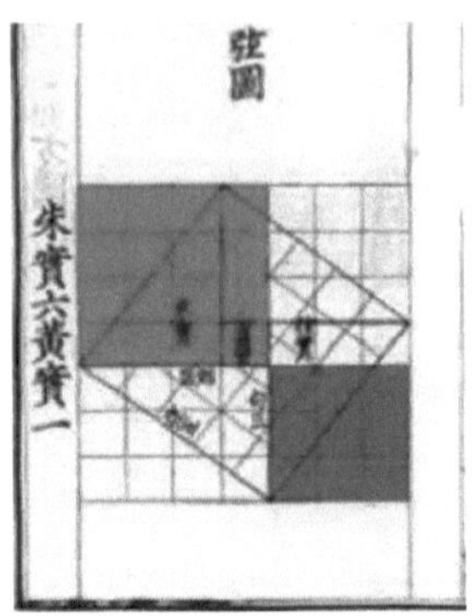

Figura 32: Diagrammatic representation of Liu's argument.
Source: Ancient China history-based task to support students' geometrical reasoning and mathematical literacy in learning Pythagoras

Fachrudin et al (2019) present a geometrical interpretation of Liu Hui's argument that is a proof of the Pythagorean theorem, which is presented below.

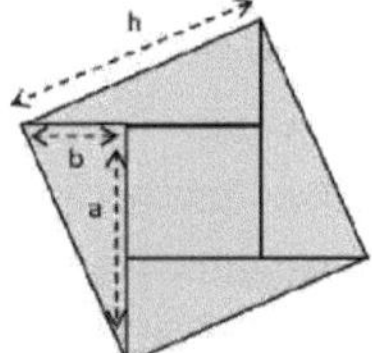

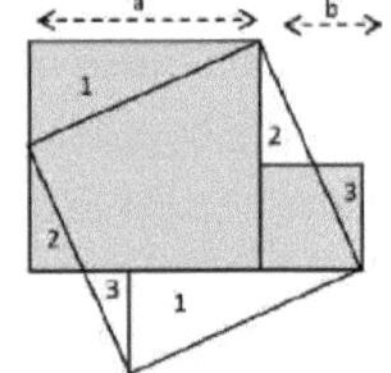

Figura 33: Diagrammatic representation of the Liu2 argument.
Source: Ancient China history-based task to support students' geometrical reasoning and mathematical literacy in learning Pythagoras

From Greece to Ancient Rome: According to the above, Greek civilisation achieved a prominent place in the history of mankind, and as far as this work is concerned, especially in mathematics. Although the Greeks absorbed and adapted elements from neighbouring civilisations, they succeeded in creating an original and outstanding culture which has managed to maintain a lasting influence on modern Western culture. The origins of Greek culture date back to about 2800 BC The Greeks migrated to a region that may have been their birthplace, known as Asia Minor, in continental Europe, what is now Greece, and in southern Italy, Sicily, Crete, Rhododara, Rosario, Madagascar and North Africa The Phoenician alphabet was introduced by the Greeks in about 775 BC, replacing the writing systems of the ancient Greeks.This led to their culture becoming more educated, which enabled them to be better able to document their history and thoughts with the help of the alphabet.

In this respect, Gil (n.d.) points out an important division of periods that frame the history of Greek mathematics, which, in his personal opinion, constitutes a

fundamental knowledge for any educator in this science.
These periods are visualised as follows:

1. The early period is called the Ionian period and runs from the end of the 7th century to the middle of the 5th century. It is characterised by the formation of mathematics as an independent science.
2. The second period lasted approximately between 450 and 300 B.C. It is called the Athenian period. Mathematics during this period of antiquity reached its full internal structure, describing what is called geometrical algebra.
3. In the third phase, the Hellenistic period, from the middle of the 4th century to the middle of the 2nd century BC. Here ancient mathematics was at its peak, especially until 150 BC. It is sometimes referred to as the Alexander period, because this is where Alexandria was undoubtedly the focal point of the ancient world's mathematical work.
4. The fourth period is characterised by mathematics in decay, a consequence of the general decline of all sciences due to decay and the collapse of the slave society; however, it is noted that important parts of ancient mathematics have survived thanks to various oriental scholars.

From the above it is necessary to point out that although the ancient Greek civilisation existed until about 600 A.D.; from the point of view of the history of mathematics, two fundamental time spans can be easily distinguished:
Classical, from 600 B.C. to 300 B.C. and Alexandrianism or Hellenism, as described above. In addition to the adoption of the alphabet and the availability of papyrus, which undoubtedly allowed the dissemination of their ideas.
Notably, Greek mathematicians excelled in their emphasis on deductive demonstrations, a unique advance compared to other civilisations. While many cultures developed basic arithmetic and geometry, only the Greeks focused on reaching conclusions through deductive reasoning. This practice contrasted radically with traditional methods of acquiring knowledge, which were based on experience, induction, analogy and experimentation. The Greeks, in their search for absolute truths, understood that deductive reasoning was the only infallible way to reach them. To ensure the soundness of their conclusions, they established axioms at the beginning of their work, allowing for a critical and well-founded examination of their foundations.
Despite their astonishing achievements, Greek mathematics still has shortcomings. Their limitations point the way to progress, but they are still not open to certain ideas. Gil (n.d.) points out that the first limitation is the inability to recognise the concept of irrational numbers. In fact, it can be assumed that there were traces in ancient Greece of what we know today as irrational numbers, which are strongly related to the idea of incommensurability

discovered by the Pythagoreans.
In this line, Arevalo (2011) argues that in modern mathematics, incommensurable reasons are expressed by irrational numbers, but the Pythagoreans would never have accepted such numbers. In fact, it was argued above that the Babylonians actually worked with such numbers as an approximation, although they probably did not know that such fractions could never be exact, just as the Egyptians did not learn to recognise the special nature of irrationals.
It is therefore interesting to note that the Pythagoreans at least understood that incommensurable reasons are of a very different kind from commensurable reasons.
It should be borne in mind that the ancient Greeks separated the study of the relations between numbers from practical arithmetic. The former was called arithmetic, while calculus was called log^stics. It is interesting to note that this division continued until the end of the 15th century, and the arithmetic of today refers to Greek log^stics, while the theory of numbers refers to the arithmetic of the Greeks. Let us recall that two segments are commensurable if it is possible to find, in a finite number of steps, a segment which serves as a common unit for the two given segments. The action of finding such a common measure gives meaning to the *verb commensurate*.
This discovery raised a central problem in Greek mathematics. Until then the Pythagoreans recognised number and geometry, but the presence of disproportionate causes destroyed this recognition. They did not stop to consider all kinds of ratios of lengths, areas and geometrical proportions, but limited themselves only to the consideration of numerical ratios. It is important to note the central idea of Pythagorean thought, in which numbers were the essence of everything, which is an integral part of the correspondence between arithmetic and geometry. Thus, in the school of Pythagoras, magnitudes began to be compared with each other in the sense of magnitude. They dealt quantitatively with the comparison of magnitudes, so that they considered every pair of magnitudes to be comparable. However, it is believed that Hippasus of Metapontus, who was part of that school (6th century BC), supposedly discovered the existence of non-commensurable quantities and calls them incommensurable quantities.
It should be noted that this discovery gave rise to what we know as irrational numbers; as a consequence, it generated a paradigmatic break with respect to the importance of geometry over arithmetic, leaving its mark on mathematical and philosophical theory.
In this line, Cardona & Munoz (2018) point out:

The Hippian revelation of the existence of unmeasurable (i.e., incommensurable) segments was, according to Merzbach & Boyer (2011), "... of profound significance for the philosophy of Pythagoras" (p. 65). According to the legend, such segments call into question all that had been established by the Pythagoreans, who based their theories on properties and relations of segments multiplied by a given unit, or on finite divisions of that unit. The unmeasurable is then presented as the unthinkable.

(p:14)

Along the same lines, Pineda & Nanez (2020) point out that for the ancient Greeks, measurement was the comparison and the process of comparing two quantities. It was known as *antiphaisresis*, and consists of making successive subtractions until a magnitude is found that manages to measure the two parts indicated, i.e., given two magnitudes, one magnitude measures the other when the first fits a natural number of times into the second. Ac- tually this is: *A* segment (magnitude) *A* measures another segment *B*, if we have that $B = nA$, for some number *n and* N.

Now, if it is given that for two magnitudes *A* and *B* this does not happen, we look for a magnitude *C* that measures them both, that is to say that $A = nC$ and $B = mC$, where *n, m and* N. This is a way of determining the number of times that *C* is in *A* and *B* respectively. If we translate this into our mathematical language we have: Two segments (magnitudes) *A* and *B* are commensurable, if there exist numbers *n* and *m* such that $nA = mB$.

We can easily see that $A = -B$, however in ancient Greek the expression - was meaningless, but it becomes meaningful with the incorporation of the *numerals*

n

rational.

Finally, with the discovery of the immeasurable, all Pythagorean demonstrations, which compared proportions of geometrical quantities, were affected and had to be reconstructed. This makes it possible to understand the sigil on the idea of irrationality on the part of the Pythagoreans and the legend of the punishment for revealing it. Legends and conjectures aside, the discovery of the irreducible quantities provoked a logical scan- dalo in all the Pythagorean circles, which was understandable, since it demanded a complete revision of its mathematical and philosophical foundations, but it was not only the cradle of Greek geometry, I consider it very necessary to make a brief review of the subject of the incommensurables, since it, as indicated, marked a turning point in the mathematical and philosophical development of ancient Greece.

Square root of two (Pythagorean constant)

Figure 34: Pythagoras Source: Taken from the Internet

He was born on the island of Samos, today's Greece, 582 B.C. and died in Metapontum, today's Italy, around 497 B.C. A Greek philosopher and mathematician, his status as the founder of a religious sect led to the early emergence of a legendary tradition around him.

The first part of his life was spent on Samos, the island he left a few years before the execution of his tyrant Pohcrates, in 522 B.C. He travelled to Miletus, and then visited Phoenicia and Egypt; in the latter country, the cradle of esoteric knowledge, he is credited with having studied the mysteries, as well as geometry and astronomy.

The community led by Pythagoras eventually became an aristocratic political force that aroused the hostility of the democratic party, leading to a revolt that forced Pythagoras to spend the last years of his life in Metapontum.

The Pythagorean community was shrouded in mystery; it seems that the disciples of Pythagoras had to wait several years before being introduced to the master and always kept strict secrecy about the teachings they received. Women could join the society; the most famous of its members was Theanus, wife of Pythagoras himself and mother of a daughter and two sons of the philosopher.

Pythagoreanism was a way of life, inspired by an austere ideal and based on the community of goods, whose main objective was the ritual purification (catharsis) of its members through the cultivation of a knowledge in which music and mathematics played an important role. The path of this knowledge was philosoria, a term which, according to tradition, Pythagoras was the first to use in its literal sense of "love of wisdom". The Pythagoreans divided scientific knowledge into four branches: Arithmetic or the science of numbers - their motto was all is number -, geometry, music and astronomy.

Numerical perfection, for the Pythagoreans, depended on the divisors of numbers. They studied properties of numbers which are familiar to us today, such as even and odd numbers, perfect numbers, friendly numbers, prime numbers, figured numbers: triangular, square, pentagonal.

The number p2

The square number of 2 was possibly the first irrational number known to the Pythagoreans. They concealed their discovery for mystical reasons. Geometrically it is the
length of the diagonal of a unit square.

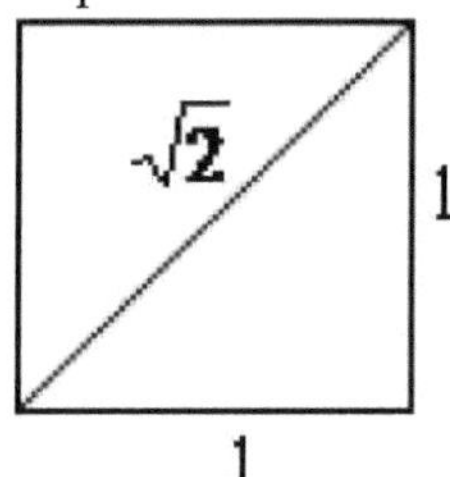

Figure 35: Number ra^z of 2 Source: Own Elaboration

The discovery of the square root of 2 as an irrational number is generally attributed to the Greek philosopher Hippos of Metapontus. He was the first to promote the demonstration of irrationality. He is said to have discovered the irrationality of the ratio of 2 when he was trying to examine a rational expression of it. However, Pythagoras believed in the unconditional definition of numbers as averages, and this forced him to disbelieve in the objectivity of irrationals. For this reason he was from the beginning against such a manifestation, for which he was sentenced to capital punishment by his Pythagorean associates.

Some representations of д/2**:** An approximation to the ralz appears in ancient India in the mathematical texts, Sulbasutras. This approximation is as follows:

1. $1+\frac{1}{3}+\frac{1}{3\cdot 4}-\frac{1}{3\cdot 4\cdot 34}=\frac{577}{408}\approx 1,414215686$

$$\sqrt{2}=1+\frac{1}{2+\frac{1}{2+\frac{1}{2+\dots}}}$$

2. As continued fractions:
3. As an infinite product:

$$\sqrt{2}=\left(1-\frac{1}{4}\right)\left(1-\frac{1}{38}\right)\left(1-\frac{1}{100}\right)\dots$$

$$\sqrt{2}=\left(\frac{2\cdot 2}{1\cdot 3}\right)\left(\frac{6\cdot 6}{5\cdot 7}\right)\left(\frac{10\cdot 10}{9\cdot 11}\right)\left(\frac{14\cdot 14}{13\cdot 15}\right)\left(\frac{18\cdot 18}{17\cdot 19}\right)\dots$$

$$\sqrt{2}=\left(1+\frac{1}{1}\right)\left(1-\frac{1}{3}\right)\left(1+\frac{1}{5}\right)\dots$$

4. With Taylor formulae for trigonometric functions :

$$\sqrt{2}=1+\frac{1}{2}-\frac{1}{2\cdot 4}-\frac{1\cdot 3}{2\cdot 4\cdot 6}-\frac{1\cdot 3\cdot 5}{2\cdot 4\cdot 6\cdot 8}\cdots$$

5. Euler methods for the series:

$$\sqrt{2}=\frac{1}{2}+\frac{3}{8}+\frac{15}{64}+\frac{35}{256}+\frac{315}{4096}+\frac{693}{16384}\cdots$$

The symbol of the square rm'z: $\surd$ was introduced in 1525 by the mathematician Christoph Rudolff to represent this operation, which appears in his book *Coss*, the first algebra treatise written in vulgar German. The sign is no more than a stylised form of the lowercase letter r to make it more elegant, lengthening it with a horizontal stroke, until it adopts its present appearance, which represents the Latin word radix, meaning rrnz. It is also conjectured that it could have arisen from the evolution of the dot that was sometimes used previously to represent it, where an oblique stroke would have been added in the direction of the radicand.

Irrationality of $\sqrt{2}$: Arithmetic demonstration of the irrationality of rrnz of 2.

It is assumed that $\sqrt{2}$ is a rational number, i.e. it can be written in the form:

$$\sqrt{2}=\frac{p}{q}$$

Under the condition that the greatest common divisor of p and q is 1.

2

Squaring and operating on the square gives: $2=\frac{p^2}{q^2}$. Therefore, the following is obtained $2p^2=q^2$

Therefore p^2 must be a multiple of 2, which implies that p is also a multiple of 2.

That is, $p = 2k$ for a certain k. Substituting, we obtain: $2q^2=(2k)^2$ y $q^2=2k^2$

So q^2 is a multiple of 2, and therefore so is q, which implies a contradiction since $(p, q) = 1$.

$\therefore \sqrt{2}$ is an irrational number

Incommensurability of the diagonal of a square with respect to its side:
Commensurability of two segments

Any two segments are commensurable if there is a common unit of measurement. For example, consider segments A and B in the figure below.

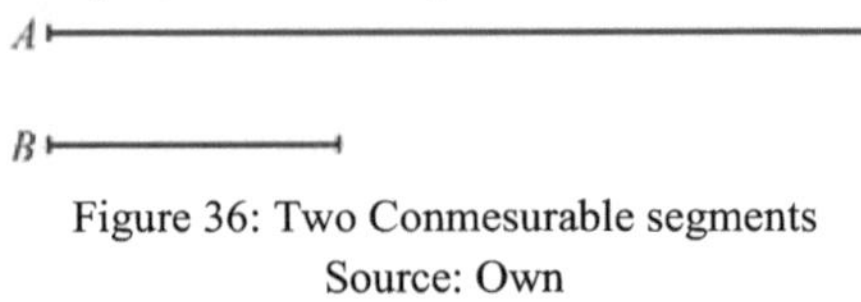

Figure 36: Two Conmesurable segments
Source: Own
Elaboration

What does it mean that two segments have a common measure?
First of all, note that segment B is smaller than segment A, so (see figure 4.9) we can include the former inside the latter as many times as it fits. This particular case shows that B fits twice inside A, but leaves a remainder: a small segment C which, naturally, is smaller than B. We can, therefore, include C inside B, as many times as it fits (in this case, one) which leaves a remainder, a segment D smaller than C. We repeat the procedure placing D inside C as many times as possible (in this case, four) and we see that there is no remainder left. Consequently, the segment D is a common measure of the segments A and B because it is contained a whole number of times in each of them: 14 times in the first one and 5 times in the second one.

The appearance of the incommensurable numbers: The appearance of the incommensurable magnitudes marked a radical change in the historical evolution of Greek geometry, since it put an end to the Pythagorean philosophical dream of number as the essence of unity, eliminated from geometry the possibility of always measuring exactly and was the first step in the development of geometry.

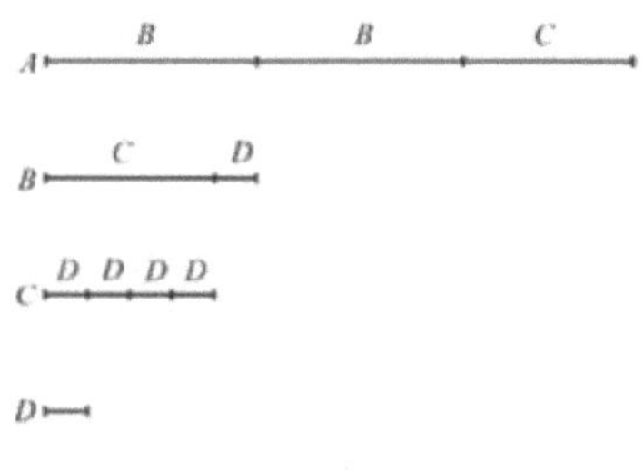

A y B están en la razón 14 : 5

Figure 37: Two Conmesurable segments 2
Source: Own Elaboration

which gave Greek mathematics a geometrical-deductive orientation embodied in the encyclopaedic compilation of Euclid's Elements. The incommensurables lead to a logical upheaval that shakes the foundations of Greek geometry, since by invalidating all the Pythagorean proofs of the theorems that used proportions they produce the first crisis of foundations in the history of mathematics.

The discovery of incommensurability marks a milestone in the history of Geometry, because it is not something empirical, but purely theoretical. Its appearance marked the most dramatic moment not only of Pythagorean Geometry but of all Greek Geometry, and it was certainly what gave Greek Mathematics a change of course that would turn it into the work of geometrical-deductive engineering embodied in Euclid's Elements.

Geometric demonstration of the incommensurability of the diagonal of a

square with respect to its side.

The demonstration of the incommensurability of the diagonal of a square with respect to its side, we will take as a reference the one carried out by Tom Apostol in the year 2000. Without loss of generality, let us consider a square of side length 1

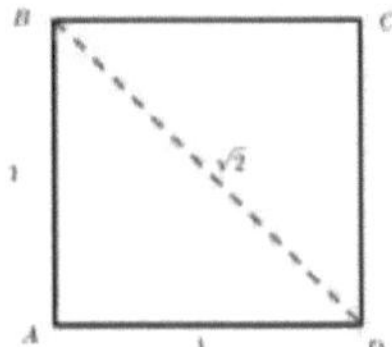

Figure 38: Square of side 1
Source: Own Elaboration

The demonstration is carried out by reductio ad absurdum. Suppose there is a common unit of measurement $l_n = 0$ which divides 1 and 2.

Consider the isosceles right triangle DAB.

With radius 1, draw an arc cutting the hypotenuse BD at a point F.

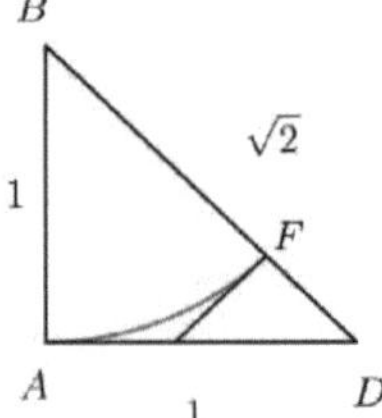

Figure 39: Demonstration 1
Source: Own elaboration

The segment FE perpendicular to BD is drawn through F.

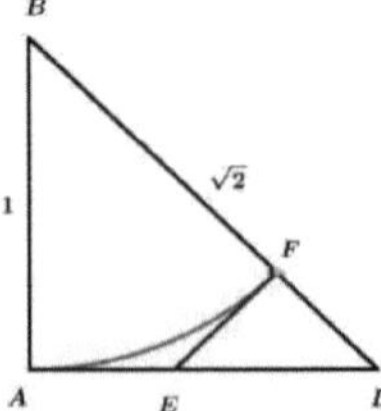

Figure 40: Demonstration 2 Source: Own elaboration

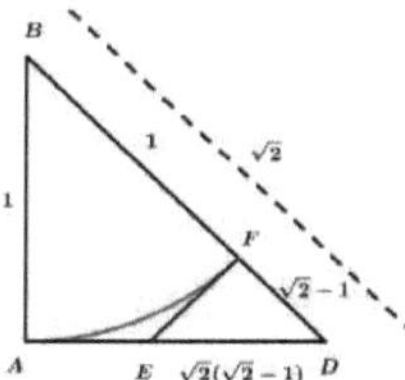

Figure 41: Demonstration 3 Source: Own Elaboration

It is evident that the triangle *EFD* is rectagular in *F* and isosceles.
Note then that,$_n$ fits *k* times in triangle *EFD* , we obtain the following data.

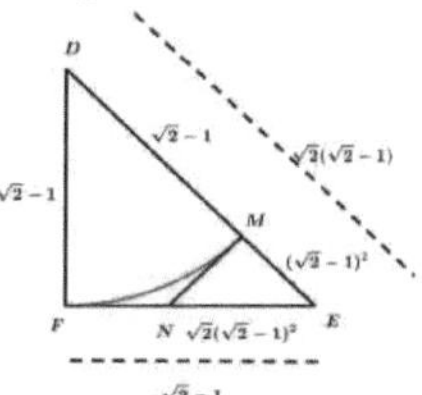

Figure 42: Demonstration 5 Source: Own elaboration

Note then that *the*,$_n$ fits *t* times in $(\sqrt{2}-1)^2$

Repeating this process, we can conclude that in the lengths of the hypotenuses and legs of the successive triangles we obtain a strictly decreasing sequence.

Iteraciones	1	2	3	4	...	n
l_n	1	$\sqrt{2}-1$	$(\sqrt{2}-1)^2$	$(\sqrt{2}-1)^3$	...	$(\sqrt{2}-1)^n$

Iterations	1	2	3	4	...	n
l_n	1	*V2* - 1	(*V2* - 1)^2	(*V2* - 1)^3	...	(*V2* - 1)^n

$l_n : \left(\sqrt{2}-1\right)^{n-1}$ applying limit to infinity, one obtains:

$$\lim_{n \to \infty} l_n = 0$$

This would imply that the segment l_n would simultaneously divide the cathetus and the hypotenuse, which leads to a contradiction, since it was assumed that $l_n \neq 0$. That is to say, these two segments are not commensurable. This leads to the historical fact that the great Pythagorean crisis occurred with the appearance of these incommensurable magnitudes. From the Pythagorean point of view, if numbers were the guiding principle of the universe, it is natural to assume that they are sufficient for the process of measuring size. It is an abstract process in which one tries to compare sizes in terms of quantity.

On the other hand, going a little into mathematics in ancient Rome, Suarez et al (2023) put forward the idea that historians of mathematics tend to exclude the ancient Romans in their work, considering them pragmatic in their application of the discipline and failing to recognise the impHicit knowledge in their great constructions. The question arises: why did they not develop mathematics theoretically, despite being well acquainted with the works of the major Greek and Hellenistic authors? It is important to note that there are few sources available on this subject, which requires a different approach to its analysis.

The ancient Romans are known to have made significant advances in engineering and architecture, creating such innovations as public fountains, communal baths, complex drainage systems, causeways and monumental buildings. Their remarkable achievements in fields such as health and infrastructure, as well as their impressive feats of Roman engineering seem to suggest advanced applications of mathematics; this idea, however, is incorrect. In fact, the Roman contribution to mathematics was quite limited.

Along these lines, Lynch 2019 notes that:

The Romans were not interested in special or logical investigation. They regularly applied simple mathematics to solve practical problems. They also needed elementary arithmetic for the supervision and management of trade and taxation, but they were satisfied with general rules which required little to understand the large body of Greek theoretical studies.

(p:1)

Roman engineers and military technicians were only interested in the simple mathematics that were essential for solving practical problems. Curiously, they had little interest in Greek trigonometry, which could have been very useful in surveying, engineering and astronomy. They ignored the beauty of theoretical mathematics and geometry that the Greeks of their time valued so highly.

Consider that by the middle of the first century BC, the Romans had consolidated their control over the ancient Greek and Hellenistic empires, which caused the mathematical development of the Greeks to grind to a halt. Although the Romans were the dominant empire on earth, no mathematical innovations occurred under their empire and within the Roman Republic, as well as no mathematicians of renown.

Interestingly, according to Lynch (2019) the Roman calendar organised by Julius Caesar had a leap year every four years in an annual cycle of 365 days which involved complex mathematical calculations. But this project was carried out by the Greek astronomer Socinius of Alexandria. His plan remained in force for 1,500 years until it was replaced in modern use by the Gregorian calendar.

On the other hand, it is said that in Rome there was a zero, but no third number. The Roman numeral zero is used only in numerical systems. Positions as decimals. The number zero does not matter. It writes the mathematical terms and operations. But the verbal ex- pressure of zero is called nulla or nihil, meaning "without" in latm.

In addition, it is known that the Romans used a counting board called "abacus", derived from the Greek, which also refers to a table or board. Initially the calculations were made with small round pebbles called calculi. Five abacuses are known from Roman times, including small bronze versions. But among the ancient Romans the word abacus had various meanings pertaining to different fields such as mathematics, architecture, home decoration, recreation, etc., although basically they all refer to objects formed by rectangular plants.

It should be noted that like ancient Greek abacuses, these Roman devices featured signs and columns to represent numbers and sums of money. However, unlike the surviving Greek examples, which are made of stone slabs, the Roman abacuses, as noted above, were made of bronze, which made them easier to transport. In addition, the fact that the fixtures were fixed in place rather than loose made them easier to use, suggesting that abacuses were widely used, although only a few actual examples survive.

On another note, Katz (2008) points out that the debate as to whether there was a "Roman mathematics" or whether all mathematical knowledge under the Roman Empire was simply an extension of "Greek mathematics" has often been discussed. Cicero (106 BC-43 BC), a prominent Roman orator, acknowledged that the Romans did not show a deep interest in mathematics, limiting it to its practical application for measuring and counting. Despite this, Cicero, as a magistrate and landowner, possessed the mathematical skills necessary to manage accounts and detect fraud. While it is true that there was no Roman equivalent of Euclid or Archimedes, the Romans employed mathematics beyond

the elementary. He claims that the Roman Empire was noted for its surveyors, who laid extensive networks of roads and aqueducts that survive to this day. However, surveying manuals of the time reveal that these professionals mainly used basic mathematical concepts. For example,
)Lucius Columella, a Roman landowner of the first century AD, emphasised the importance of knowledge of area calculation for those working in agriculture, providing simple formulas for calculating areas of figures(such as squares, rectangles, triangles and circles. In addition, he described methods such as $\left(\frac{1}{3}+\frac{1}{9}\right)\cdot s^2$ to calculate the area of an equilateral triangle with side *s*, and used $\frac{22}{7}$ as an approximation of . π.
However, he points out that Vitruvius, a first century B.C. writer, a contemporary of Julius Caesar and Augustus, is an example of someone with a sound knowledge of mathematics. In his work, On Architecture, he insisted that architects should receive a broad education ranging from technical drawing to astronomy. Vitruvius stressed the importance of geometry in architecture, for the use of tools such as the compass and the ruler, and arithmetic for calculating building costs. He also explained practical methods such as determining true north using a gnomon and solving symmetry problems with geometrical principles. Although he recommended mathematical knowledge, he only gave basic examples of architecture, without going into advanced mathematical concepts.
Cuomo (2021) also refers to Vitruvius, who identifies him not only as a writer, but also as an architect and military engineer. In his only work, "La Arquitectura", he covers various topics, highlighting machines and sundials in book 10. In this work, mathematics plays a crucial role in his work, as he uses it to draw plans of buildings, determine the direction of the winds, calculate the proportions of the elements of a temple from a standard module, construct resonator vessels to amplify voices in theatres according to the principles of the armorna, and elaborate an analemma, which is the basis of sundials. In addition, Vitruvius provides pre-established measurements for catapults, calculated according to the weight of the projectile, to help those unfamiliar with geometrical procedures to obtain this information quickly during a siege.
This same author, who also refers to Marcus Junius Nipsus (2nd century A.D.), who was a Roman grammatical writer who also dealt with various mathematical questions, also makes reference to the role of mathematics in the Roman Empire. His surviving writings include *the Corpus Agrimensorum Romanorum*, a compilation of Latin works on surveying. For Cuomo (2021), the

importance of Marcus Junius Nipsus is attributed to the fact that he is a great devotee of the surveyor's mathematical knowledge. In his work "Measurement of a surface", we can appreciate definitions of measures and angles, followed by a series of problems, most of them about triangles such as "given an odd number, form a right triangle", "given an even number, form a right triangle". In this respect, we have the following excerpt from Marcus Junius Nipsus' work: measure the area of all triangles by a single method, say, rectangle, acute-angle and obtusangle. We would cover it in this way. I put the three numbers of any of the three triangles together. That is, the rectangle, whose numbers are given, the cathetus 6 feet, the base 8 feet, the hy- potenuse 10 feet, I put these three numbers into one, and they add up to 24. It comes out 12. This I set aside, and from this number, that is, from 12, I subtract the *(other)* numbers in- dividual. Subtract 6: I put the remainder under 12. In an analogous way I subtract the base, 8 feet, from 12: I put the remainder under 6. Then I multiply 6 by 4. I multiply this by 2, and it gives 48. This I multiply by 12. Gives 576. From this I take the nn'z, and it gives 24. That will be the area. And the area of the other triangles will be calculated in the same way.

(Cuomo (2021): 171)

Conclusions

In this fascinating exploration of the history and origins of numbers, we briefly review how various number systems have influenced human development. From the impressive Egyptian hieroglyphic system to the advanced Babylonian sexagesimal system, to complex Hindu and Chinese resolutions, each method illustrates the creativity and adaptability of ancient civilisations in the face of everyday challenges. For example, the Egyptians created a decimal system using hieroglyphs to represent numbers such as ten, which facilitated accounting and the construction of their great buildings. On the other hand, the Babylonians employed a sexagesimal system, the basis of our current measures of time and angles, and excelled in the use of fractions.

As for the Romans, although their mathematical contribution was not as theoretical as that of the Greeks, their practical approach was crucial. The Romans were expert surveyors who used basic mathematics to manage their vast empire. They developed simple but effective methods for measuring land, calculating areas and building infrastructures such as roads and aqueducts. Roman abacuses, made of bronze with fixed counters, were practical tools for calculations and records in commerce and administration. Although Roman surveying manuals used relatively simple methods, such as those described by Marcus Junius Nipsus for measuring rivers, they were adequate for the needs of their time. This shows that, despite not having an advanced mathematical theory, the Romans integrated mathematics into their daily life in a very functional way.

An equally remarkable aspect is ancient Chinese mathematics, which dates back more than 4,000 years and displays a richness and originality of its own. The ancient Chinese mathematicians made significant contributions in arithmetic" (The Nine Books of Mathematics), evidencing a deep and practical understanding of mathematics, including advanced concepts such as the theory of equations and the rule of three. Furthermore, the Chinese were already familiar with the Pythagorean theorem long before it was formulated, describing the properties of right-angled triangles. The ancient Chinese also developed methods for calculating square roots with remarkable accuracy, using iterative procedures that anticipated later techniques.

Understanding the history of numbers and number systems helps educators to teach students the importance of flexibility and innovation in mathematical problem solving. This historical perspective also reveals how different cultures, including Chinese, have contributed to the mathematical knowledge we take for granted today. This understanding allows teachers to underline the relevance of concepts such as decimal approximations, the value of n and square ratios, which are fundamental not only in mathematical theory, but also in practical applications in engineering and science. In particular, the value of n has been the subject of study and research for centuries, from the earliest approximations in Egypt and Babylon to more precise estimates in later times. Thus educators can encourage students to appreciate both the history and current applications of mathematics, developing an understanding that goes beyond the textbook material.

Although this analysis has briefly touched on the richness of Greek mathematics, it is important to recognise that its depth and breadth deserve more detailed attention. The Greeks made significant contributions with leading figures such as Euclid, Pythagoras and Archimedes laying the foundations for many branches of modern mathematics. From geometry to number theory, the influence of Greek mathematics is extensive and significant. It has therefore been decided to reserve a more exhaustive analysis of the Greek contributions for future publications, in order to better understand the impact and legacy of these pioneers of mathematical thought.

In short, the study of the history of numbers and counting systems not only offers insight into human cultural and technological development, but is also crucial for contemporary mathematics education. This knowledge helps to understand how mathematical thinking has become a universal constant, adapting and evolving in different historical and geographical contexts. By incorporating this perspective into teaching, students acquire not only mathematical skills, but also a deep respect for humanity's intellectual legacy. This historical and contextual view enriches learning and fosters a lasting understanding of the beauty and utility of mathematics.

For a mathematics teacher, knowing the evolution and historical impact of numerical and mathematical systems not only enriches their own understanding of the subject, but also provides tools for teaching with greater depth and context. By integrating the history of mathematics into teaching, teachers can help students connect abstract concepts with their historical and cultural applications, promoting a deeper appreciation of the subject. Furthermore, this knowledge allows educators to present mathematics not only as a set of techniques and procedures, but as a living and evolving discipline, highlighting how mathematical ideas have influenced human development over time.

Bibliography

[1] Arevalo, N (2011). Analisis historico-epistemologico del concepto de numero irra- cional y los obstaculos presentes en su transposition textual (licenciatura). Univer- sidad del valle. Https://core.ac.uk/download/pdf/157765269.pdf

[2] Beckmann, P. (2006). *A history of n*. Mexico: Libraria, SA de CV.

[3] Berciano, A. (2007). *Matematicas en elAntiguo Egipto*. Vasco.

[4] Caicedo, D., & Madrigal, G. (2017). *La historia de la matematica como recurso didactico y alternativa de aprendizaje de los numeros irracionales* [Unpublished master's thesis]. University of Medellm. https://repository.udem.edu.co/ bitstream/handle/11407/4653/T_MEM_48.pdf

[5] Cardona, S., & Munoz, J. (2018). Episodes of the adventure of the irrational in the dawn of modernity [Master's Thesis, University of Medellm]. https://repository.udem.edu.co/bitstream/handle/11407/6292/T_MEM_354.pdf?sequence =2&isAllowed=y

[6] Chaves, E., & Salazar, J. (2003). El papel y algunas condiciones para la utiliza- tion de la Historia de la Matematica como recurso metodologico en los procesos de ensenanza-aprendizaje de la Matematica. http://cimm.ucr.ac.cr/ojs/index. php/eudoxus/article/download/109/102

[7] Corrales, J. (2022). *Un paseo por el maravilloso mundo de los numeros: La historia como recurso didactico en matematicas* [Universidad de Alca la]. https://ebuah.uah.es/dspace/bitstream/handle/10017/53952/TFM_ Corrales_Martinez_2022.pdf?sequence=1&isAllowed=y

[8] Crespo, C. (2008). Acerca de la comprension y significado de los numeros irracio- nales en el aula de matematica. http://www.soarem.org.ar/Documentos/41% 20Crespo.pdf

[9] Cuomo, S. (2001). Ancient Mathematics. Routledge. https://doi.org/10.4324/9780203995730

[10] Everett, C. (2021). *The numbers made us the way we are*. Critica, Barcelona.

[11] Eves, H. (1964). *An Introduction to the History of Mathematics*. New York: Rinehart & Winston.

[12] Fachrudin, A.et al (2019). Ancient China history-based task to support students' geometrical reasoning and mathematical literacy in learning Pythagoras. Journal of Physics Conference Series, 1417(1), 012042. https://doi.org/10.1088/1742-6596/1417/1/012042.

[13] Fernandez, E. (2010). Babylon and mathematics in the classroom. *Revista Digital de Ciencias Bezmiliana*. https://www.clubcientificobezmiliana.org/

revista/images/stories/babiloniamatematicasaula.pdf

[14] Garda, G. (2009). Notes for an interpretation of Heron de Alejan- dria's method. https://semana.mat.uson.mx/semanaxxvii/Memorias/XIX.pdf .

[15] Gervan, H. (2015). Mathematical practice in Ancient Egypt: A re-reading of Problem 10 of the Moscow Mathematical Papyrus. *Edu.ar*. https://revistas.unc. edu.ar/index.php/anuariohistoria/article/view/12511/12787

[16] Gil, M. (n.d.). The decline of Hellenistic mathematics and mathematics in Rome. University of Castilla-La Mancha https://matematicas.uclm.es/itacr/web_matematicas/ papers/3/3_cases_math_helena_helena.pdf

[17] Gillings, R. (1982). *Mathematics in the Time of the Pharaohs*. Courier Corporation.

[18] Gonzalez, F., Martm-Loeches, M., & Pobes, E. (2010). Prehistory of mathematics and the modern mind: Mathematical thinking and recursion in the Franco-Cantabrian Palaeohistory. *Dynamis, 30*. https://doi.org/10.4321/ s0211-95362010000100007

[19] Gonzalez, P. (2004). La historia de las matematicas como recurso didactico e ins- trumento para enriquecer culturalmente su ensenanza. http://www.revistasuma. es/index.php?option=com_docman&task=cat_view&gid=5&limitstart= 5

[20] Guzman, M. (2007). Ensenanza de las ciencias y la matematica. http://www. rieoei.org/rie43a02.pdf

[21] Ifrah, G. (1987). *From One to Zero: A Universal History of Numbers*. Penguin Books.

[22] Katz, V. (2008). A history of mathematics (3rd ed.). Pearson.

[23] Lupianez, J. (2009). Historia de la Ensenanza de las Matematicas. http://cimm. ucr.ac.cr/ojs/index.php/eudoxus/article/view/119

[24] Lynch, P. (2019, July 4). What did the Romans ever do for maths? Very little. The Irish Times. https://www.irishtimes.com/news/science/what-did-the-romans- ever-do-for-maths-very-little-1.3940438

[25] Martmez, A. (2001). El diseno de piramides basadas en el triangulo sagrado egip- cio. *Boletm de la Asociación Espanola de Egiplologi'a, 11*, 7-20. https://dialnet. unirioja.es/servlet/articulo?codigo=2613392

[26] Mayoral, C. (2009). *Propuesta metodologica para la aproximacion de rcu'ces cuadradas, cubicas y quintas* [Tesis de maestri'a|. Universi- dation Tecnologica de Pereira. https://repositorio.utp.edu.co/items/ d033574d-8366-41a6-9c13-b9dc72037715

[27] Miralles de Imperial, J., & Deulofeu, J. (2005). Historia y ensenanza de la matematica: Aproximaciones de las nn'ces cuadradas. *Educacion Matematica,*

17(1), 87-106.

[28] Morales, L. (2002). Las Matematicas en el Antiguo Egipto. http://euler.mat. uson.mx/depto/publicaciones/apuntes/pdf/1-1-1-egipto.pdf

[29] Moreno, R. (2012). *Las Matematicas de los faraones*. Nivola Libros y Ediciones, S.L.

[30] Ortiz, A. (2005). *Historia de la Matematica. Volume 1: La Matematica en la Antiguedad*. Edu.pe. https://textos.pucp.edu.pe/pdf/2389.pdf

[31] Pineda, D. & Nanez, Y. (2020). Historical-epistemological development of irrational numbers. In Sigma eBooks (Vol. 16, Number 1, pp. 33-49). https://dialnet.unirioja.es/descarga/articulo/7667725.pdf

[32] Plato (1990). Theaetetus (Balasch, M.). Editorial Anthropos (original work published in the 4th century BC).

[33] Pena, E. (2013). *Historia de Numeros*. Umag.cl. Retrieved February 2, 2024, from https://kataix.umag.cl/~edopena/material/apuntes/

[34] Ribnikov, K.(1987). History of mathematics. Moscow, Mir.

[35] Romero, J., et al. (2021). Analysis of algebraic equations from Egyptian culture to the present day. *Zenodo (CERN European Organization for Nuclear Research)*. https://doi.org/10.5281/zenodo.5500693

[36] Sanchez, M. (2012). Pitagoras, el teorema de Pitagoras: un secreto encerra- do en tres paredes. https://vivelacienciacom.wordpress.com/wp-content/uploads/2020/05/13gic-pitagoras.pdf

[37] Sirotic, N., & Zazkis, R. (2004). Making Sense of Irrational Numbers: Focusing on Representation. In *Proceedings of the 28th International Conference for the Psychology of Mathematics Education*, Bergen, Norway, Vol. 4, pp. 497-505.

[38] Sirotic, N., & Zazkis, R. (2007). Irrational Numbers on the Number Line-Where Are They? *International Journal of Mathematical Education in Science and Technology, 38*(4), 477-488.

[39] Sirotic, N., & Zazkis, R. (2010). Representing and defining irrational numbers: Exposing the missing link. *CBMS Issues in Mathematics Education*.

[40] Stewart, I. (2008). *History of mathematics: In the last 1000 years*. Gru- po Planeta (GBS). https://www.tomasdeaquino.cl/upfiles/documentos/31072018_853am_5b60780498062.pdf

[41] Suarez, J, Davila, & Esquivel, A. (2023). Mathematics in ancient Rome, through the study of ethnomathematics. Educateconciencia, 31(40), 101-126. *https* : //doi.org/10,58299/edu.v31i40,682

[42] Urbaneja, P. (2004). La historia de las matematicas como recurso didactico e ins- trumento para enriquecer culturalmente su ensenanza. *Suma:*

Journal on Teaching and Learning Mathematics, 45, 17-28. http://funes.uniandes.edu. co/7239/.

Printed by Books on Demand GmbH, Norderstedt / Germany